Green Buildings and the Law

In countries such as the UK, the energy used in constructing, occupying and operating buildings represents approximately 50 per cent of greenhouse gas emissions. Pressure to improve the environmental performance of buildings during both construction and occupancy, particularly to reduce carbon emissions from buildings, has become intense.

Understandably, legislation and regulation are driving green development and compliance. And this is happening in a wide variety of ways. This review of the law in key jurisdictions for the research community, lawyers, the construction industry and government examines some of the mechanisms in place – from the more traditional building regulation controls to green leases and the law relating to buildings and their natural environment. Members of the CIB TG69 research group on 'Green Buildings and the Law' review aspects of the law relating to green development in a range of jurisdictions.

Julie Adshead is Associate Dean (Enterprise) and Senior Lecturer at the University of Salford Law School, UK and co-ordinator of the CIB TG69 Research Group on 'Green Buildings and the Law'.

About CIB and about the series

CIB, the International Council for Research and Innovation in Building and Construction, was established in 1953 to stimulate and facilitate international cooperation and information exchange between governmental research institutes in the building and construction sector, with an emphasis on those institutes engaged in technical fields of research.

CIB has since developed into a world-wide network of over 5000 experts from about 500 member organisations active in the research community, in industry or in education, who cooperate and exchange information in over 50 CIB Commissions and Task Groups covering all fields in building and construction related research and innovation.

http://www.cibworld.nl/

This series consists of a careful selection of state-of-the-art reports and conference proceedings from CIB activities.

Open & Industrialized Building *A Sarja*
ISBN: 0419238409. Published: 1998

Building Education and Research proceedings *J Yang* et al.
ISBN: 041923800X. Published: 1998

Dispute Resolution and Conflict Management *P Fenn* et al.
ISBN: 0419237003. Published: 1998

Profitable Partnering in Construction *S Ogunlana*
ISBN: 0419247602. Published: 1999

Case Studies in Post-Construction Liability *A Lavers*
ISBN: 0419245707. Published: 1999

Cost Modelling *M Skitmore* et al. (allied series: Foundation of the Built Environment)
ISBN: 0419192301. Published: 1999

Procurement Systems *S Rowlinson* et al.
ISBN: 0419241000. Published: 1999

Residential Open Building *S Kendall* et al.
ISBN: 0419238301. Published: 1999

Innovation in Construction *A Manseau* et al.
ISBN: 0415254787. Published: 2001

Construction Safety Management Systems *S Rowlinson*
ISBN: 0415300630. Published: 2004

Response Control and Seismic Isolation of Buildings *M Higashino* et al.
ISBN: 0415366232. Published: 2006

Mediation in the Construction Industry *P. Brooker* et al.
ISBN: 0415471753. Published: 2010

Green Buildings and the Law *J. Adshead*
ISBN: 90415559263. Published: 2011

Green Buildings and the Law

Edited by Julie Adshead

LONDON AND NEW YORK

First published 2011 by SPON Press

2 Park Square, Milton Park, Abingdon, Oxfordshire OX14 4RN
52 Vanderbilt Avenue, New York, NY 10017

Routledge is an imprint of the Taylor & Francis Group, an informa business

First issued in paperback 2019

British Library Cataloguing in Publication Data
A catalogue record for this book is available from the British Library

Library of Congress Cataloging-in-Publication Data
A catalog record has been requested for this book

ISBN13: 978-0-415-55926-3 (hbk)
ISBN13: 978-0-367-86549-8 (pbk)

Typeset in Sabon by
Keystroke, Station Road, Wolverhampton

Contents

Dr Jeremy Gibberd is an architect with research interests in sustainable built environments and educational buildings. He provides technical advice in these areas to government and the private sector and has worked the UK, the USA and South Africa.

Dr Asanga Gunawansa holds a Ph.D. in law from the National University of Singapore (NUS) and an LL.M. in International Economic Law from the University of Warwick. He is an Attorney-at-Law of the Supreme Court of Sri Lanka. He is currently attached to the Department of Building of the National University of Singapore as an assistant professor. He is also a Research Associate of the Institute of Water Policy of the Lee Kuan Yew School of Public Policy and an Associate Member of the Executive Committee of the Asia Pacific Centre of Environmental Law. His teaching and research areas include construction law, ADR, legal aspects of infrastructure development, international environmental law, and legal aspects of sustainable development. Prior to joining NUS in May 2007, he worked as a legal officer for the United Nations Organization for over seven years. During the period 1993–2000, he was a State Counsel attached to the Attorney General's Department of Sri Lanka.

Dr Deniz Ilter received her B.Arch., M.Arch and Ph.D. degrees from Istanbul Technical University. She is currently a member in ITU's Civil Engineering Department and is an accredited BREEAM Assessor.

Dr Andrew H. Kelly is Associate Professor at the University of Wollongong, NSW, Australia, where he is also a member of the Institute of Biological Conservation and Environmental Management. He is qualified in both town planning and law, with research interests in local government, biodiversity conservation and the rural/urban interface.

Robert A. Leiter, FAICP, an urban planner for more than 35 years, has served as Planning Director for several California cities and recently retired as Planning Director for the San Diego Association of Governments. Mr. Leiter holds a BA in political science and MA in economics from the University of California at Santa Barbara. He was elected to the College of Fellows of the American Institute of Certified Planners in 2007, and is Chair of the American Planning Association's Regional and Intergovernmental Planning Division.

Stuart J. Little is employed as the Rural Lands Program Coordinator with the Sydney Catchment Authority. From 1992 to 2004, he worked as a Senior Environmental Planner with the NSW Department of Planning, focusing on urban development, biodiversity and bushfire policy issues. He continues to maintain an active research interest in these areas.

Tamera L. McCuen is an Assistant Professor in the Haskell and Irene Lemon Construction Science Division in the College of Architecture at the University of Oklahoma. Her research includes an investigation of the relationships between sustainable, high-performance buildings and project delivery methods.

Lorraine Murphy MSc, AIEMA is a researcher at OTB Research Institute for the Built Environment in the Netherlands. Her research interests include the functioning of 'energy'-related policy instruments in buildings.

Rui Guan Michael is Associate Director, Ho & Partners Architects, Engineers & Development Consultants Ltd. His research interests include building regulations in Asia, (Hong Kong, Mainland China and Singapore); green building and building controls; China's housing policy, building laws and codes; comparative study on architects' professional practice (China, Hong Kong and Singapore); architecture conservation and building controls; and China's architectural history and theory (colonial architecture, Manchuria area).

Alfred A. Talukhaba holds a Ph.D. in the field of Construction Management. He is an associate professor at the School of Construction Economics and Management at the University of the Witwatersrand, South Africa. He heads the Research Group on Energy Efficiency and lectures building science to undergraduate students as well as supervising masters and doctorate students in energy efficiency in buildings. He is a member of the Royal Institution of Charted Surveyors (RICS), a member of the Chartered Institute of Building (CIOB) and an associate member of the Chartered Institute of Arbitration (CIArb). Currently, he is the Senior Vice President of CIOB Africa.

Colleen Theron is a solicitor and environmental consultant specializing in environmental law and sustainability issues. She is the director of CLT Envirolaw and is a member of the Lexis Legal Intelligence team. Colleen also lectures at Birkbeck College in London. She works widely with business on environmental compliance, risk and sustainability issues. Colleen is recognized as a leader in her field in both Chambers and Legal 500. She is a frequent contributor to legal and professional journals and is an executive member of the UK Environmental Law Association.

Joachim E. Wafula holds an Honours degree in Quantity Surveying/ Construction Economics and is on course towards his MSc (Building) qualification at the University of the Witwatersrand. His research interests are in Energy Efficiency in building regulations, carbon trading in the energy sector in emerging markets and sustainability in the built environment.

Preface

The formation of Working Commission W113 in 2006 signalled the emergence of a substantial body of legal research within CIB. Increasingly, environmental lawyers were drawn to this group. The significant environmental impact of urbanization and construction activity was well recognized, as was the consequent relevance of environmental law to the construction industry and to CIB.

Against the backdrop of legislative developments responding to the global warming imperative and the Kyoto Agreement and calls for increasingly more stringent targets for carbon dioxide emissions, a group of legal researchers from both within and outside the W113 Working Commission came together to establish a world-wide project looking at the law relating to green buildings. The impact of construction and building occupancy upon greenhouse gas emissions was well documented, with energy use in these areas contributing in the UK, for example, to around 50 per cent of the total emissions. However, aside from the need to reduce carbon emissions across the globe, this group of legal scholars recognized an overarching necessity to consider in a holistic way how buildings can have an adverse or positive impact upon their surrounding environment. This fundamental question is of particular importance to the construction industry as well as the international construction research community.

TG69 on Green Buildings and the Law was established as a CIB Task Group in 2007 under the umbrella of the W113 Working Commission. Recognizing that there was a complex and multi-layered body of legislation on green buildings in place world wide, the remit of the Task Group was to review the law (as well as soft law and voluntary mechanisms) across key jurisdictions and thus provide the research community, lawyers and industry with important information about the legal situation in these areas across the world. The need for international co-ordination was self-evident. Global warming and environmental degradation present common problems and it is crucial that best practice is shared and knowledge of the existing and prospective legal regimes is disseminated world wide.

The ultimate goal of the TG69 Task Group was to produce a CIB publication containing a selection of research outputs authored by prominent environmental law scholars across the world. This current publication sees the culmination of the research project and seeks to provide an outline of the state of the law relating to green buildings in key jurisdictions and address some of the legal challenges presented by the global warming imperative. It is hoped that important lessons can be drawn from this work by lawyers, legislators and government alike and that individual nations will be able to learn from the experience of others who are at different stages in the process of introducing legal controls.

Julie Adshead
November 2010

Introduction

Julie Adshead

In adopting a holistic approach to the interaction between buildings and their environments, it soon became clear to the research group that there is a wide range of different ways in which buildings can impact upon the environment (and vice versa) and, consequently, many different areas of law to consider. Although the main focus of the study is upon standards for sustainable buildings, such wide-ranging areas as constitutional law, planning law and product standards are also considered. There are eighteen individual contributors to the twelve chapters of the book. Snapshots of the law relating to aspects of green building are presented from nine different jurisdictions. The task of determining an order for these contributions did not prove to be an easy one. Ultimately, the chapters have been grouped into four sections; the first contains two contributions from the United States, the second contains four chapters from Europe, the third has two chapters from contributors in South Africa and the final section contains four chapters from Asia and Australia.

The United States

Neither of the two opening chapters from the United States takes, what might be considered, the traditional approach to green building law. There tends to be an assumption that the law in this area is all about controlling emissions from buildings by putting in place standards to be met for their design and function. Buildings also have an effect upon the environment through the materials and processes that are used in their construction and the planning and regulatory framework that determines their nature and location. The first contribution from the United States is from Robert A. Leiter. Now retired, Robert was an urban planner for over 35 years. He served as planning director for several California cities and most recently held the position of Planning Director for the San Diego Association of Governments. This opening chapter charts the course of the inclusion of environmental and sustainability considerations in the US planning and regulatory framework over the past 40 years. The focus is upon planning and policy approaches in San Diego, California, and a case study is used to illustrate the evolutionary process. Environmental systems planning and comprehensive planning have very much become features in planning regimes and Leiter concludes that, in the United States, the federal and state laws currently in place, provide a framework that allows states to address regional and local community issues. The theme of central control as opposed to local control is a recurrent one (discussed below) and there are some interesting comparisons to be drawn from this chapter with the ways in which the planning regime in the UK has increasingly embraced environmental considerations and sustainability goals.

The second chapter from the United States is authored by Tamera L. McCuen and Lee A. Fithian, who are both senior academics at the University of Oklahoma. McCuen and Fithian explore some of the inconsistencies that exist between what may, on the face of it, appear to be a 'green building' and the underlying environmental consequences of certain materials, processes and practices, as well as the occupancy phase of a building's life. What may seem to be a 'green' building material may involve environmentally damaging processes, result in hazardous by-products and present challenges for recycling and reuse and the construction waste stream. Construction equipment used on site can also be responsible for the production of deadly toxins. Reduced energy consumption of a building during occupancy is a key element of 'green building' and yet there are environmentally damaging consequences to be considered, even with such widely advocated alternative energy sources as ground source heat pumps. McCuen and Fithian identify a challenge in the US that exists worldwide: the divergence between the commitment to achieving 'green building' on the one hand and the wide range of current standards, codes, regulations and governance processes in place that do not always fully support that goal. This chapter delivers a clear message with regard to the need for a holistic 'cradle to grave' approach when considering buildings and the environment.

The European Union

There are three chapters originating from EU member states; two from the United Kingdom and one from the Netherlands. Also included in this section is a chapter from Turkey. Turkey is not currently a member of the EU, but has adopted much of the body of EU law, referred to as the '*acquis*'. Although there is now quite a substantial quantity of EU law in place governing emissions from buildings and products, it is, nonetheless, possible to identify quite different initiatives and levels of progress across the Union. The author of the Netherlands chapter is Lorraine Murphy, a researcher at the OTB Research Institute for the Built Environment. The institute is part of the Technical University of Delft but also conducts independent research and consultancy in the areas of housing, construction and the built environment. The OTB Research Institute is recognized as one of the world leaders in research on building control and regulations. The Netherlands is often considered to be a forerunner in the field of sustainable or green buildings, yet Murphy notes that green buildings are still not 'mainstream', particularly where existing stock is concerned. It is also interesting that, after long debate over how sustainability should be measured, the environmental chapter of the Netherlands National Building Decree still remains empty. The contribution from the Netherlands considers the combined role of building regulations and voluntary agreements in driving up standards for building performance. The Netherlands commonly uses voluntary agreements or covenants to supplement the legal framework for many sector-specific issues and has recently introduced several to promote energy performance improvement in buildings. The covenant is a voluntary consensual agreement between government and sector representatives over a set period of time. There has been much debate over the merits of regulation as opposed to voluntary measures and this forms another theme across the chapters. Murphy sees possibilities in a combined approach. It is too early to judge the performance of the recent building sector covenants in concrete terms, but she notes that they show promise and could serve as a useful test-bed for subsequent legislation.

The use of voluntary mechanisms to achieve higher standards of sustainability in buildings is also considered in the first of the chapters from the United Kingdom. The focus of this contribution is upon how the planning regime can serve to help achieve improvements in building sustainability. A voluntary code 'The Code for Sustainable Buildings' has been operational since 2006 in the UK. Although still fundamentally voluntary in nature, since 2008 it became necessary for new buildings to either have a certificate of rating under the code or a statement of non-assessment. At the same time, adherence to level 3 of the code became mandatory for all new public buildings. Previously the lowest rated level of performance under the code was set above the minimum standard required in building regulations. Local planning authorities, therefore, used conditions in their planning determinations to require new residential housing to meet higher standards as set out in the Code. It appears likely, however, that new government guidance will severely limit the degree to which the planning system in the UK can be used to encourage higher standards and the use of on-site renewables. The question of central as opposed to local control arises again in this chapter. It seems that a single central control mechanism in the form of building regulations is currently the preferred option in the UK.

Deniz Ilter, the author of the Turkish chapter, is an academic at the Istanbul Technical University. In order to assess the legislative provisions in place in Turkey to support green building, Ilter identifies five green building themes: energy, water, materials, site and waste. The survey of Turkish legislation reveals that energy efficiency is the most advanced of these themes, with well-established legislation in place. There is very little by way of control in terms of water efficiency in Turkey, which is noted to be a serious defect in an environment already subject to water shortage and in light of further threats to water supply in parallel with global warming. Ilter is the second contributor who considers the building materials used in construction and their impact upon the environment. In terms of the controls in place in this regard in Turkey, she concludes that further legislative measures are needed. Waste is another theme that is subject to well-established legislation and is well controlled in this jurisdiction. Site location is, however, in the author's view, an issue that warrants further consideration in Turkey. Aside from the need to develop the legislative provisions in several of the thematic areas, Ilter recognizes that there are underpinning problems in Turkey in relation to resources and education as well as with institutional organization. The situation in Turkey provides an interesting comparison to that in the Netherlands. As a potential candidate member of the EU, Turkey has adopted most of the Union's legislation. However, there is a stark contrast between the level of development of sustainability provisions between Turkey and the Netherlands, the latter member state being generally at the forefront of 'green' developments in the EU.

Colleen Theron and Malcolm Dowden are both prominent UK environmental lawyers. They are currently with LexisNexis, the legal information provider. Theron and Dowden co-author the second contribution from the United Kingdom. This chapter considers the impact of green buildings upon the commercial real estate sector and *vice versa*. There is little question that standards of sustainability in buildings would be driven up if a tangible financial benefit to green building could be recognized. Theron and Dowden look at some of the drivers such as increased rental potential and lower operating costs. They also consider the use of green leases to encourage or require landlords and tenants to manage their assets sustainably and the impact of the recently introduced carbon reduction commitment energy efficiency scheme (CRC). The CRC is a mandatory emissions trading scheme for the UK, designed to reduce carbon dioxide emissions from commercial and publicly owned premises. Poor performance under the scheme could have a significant impact in terms of both financial costs and reputation for building owners. The authors conclude that the real estate sector will have to embrace energy efficiency in order to meet corporate social responsibility requirements, confront rising energy prices, comply with legislative requirements and remain competitive. However, they recognize a need for greater certainty, both in terms of what sustainability actually means and what the important factors to tenants and occupiers are, before improved building sustainability is converted to higher valuation. This returns us to the fundamental problem of defining and measuring the 'sustainability' or 'greenness' of building and, indeed, whether the two concepts are the same or completely different.

South Africa

South Africa is currently facing many challenges, not least of which is the level of its carbon emissions. Carbon emissions here are the highest in Africa and South Africa has a high energy consumption per capita due to energy used in mines and manufacturing, 90 per cent of which is produced through coal-powered stations. Jeremy Gibberd authors the first of two chapters from South Africa. He is an architect who provides technical advice on sustainable built environments to government and the private sector. This chapter goes back to the very basics of legislation in examining whether the South African Constitution provides a framework that will transfer to national legislation to promote sustainable building. For most jurisdictions this is not a question that arises as older constitutions do not reflect the principle of sustainable development at all. However, the South African Constitution, adopted in 1996, does contain provisions on sustainability and the author concludes that it is an appropriate overarching legal framework to address sustainable development. Nonetheless, current building-related legislation in South Africa, in his opinion, does not reflect the requirements of the Constitution in this regard. In particular, the health and well-being provisions in the Constitution require further development in legislation and indigenous, sustainable construction needs to be recognized and supported in building regulations. Gibberd further concludes that in order to reflect the sustainable development requirements of the Constitution, there need to be mandatory requirements, for example for water and energy efficiency, in national legislation. The failure to promote and recognize sustainable indigenous construction practices is an issue that also arises in the Chinese chapter below. This perhaps reflects a tendency to assume that the most modern and technologically developed techniques will yield the best results in terms of sustainability and efficiency. This is, of course, not necessarily always the case.

The second chapter from South Africa focuses upon building regulations and control. The authors, Associate Professor Alfred A. Talukhaba, Joachim E. Wafula and Kennedy O. Aduda are all from the University of Witwatersrand in Johannesburg. The need to improve energy efficiency has recently been driven up the agenda in South Africa by a 32 per cent increase in energy costs in 2008 and planned increases of around 25 per cent per annum over the next three years. Security of energy supply also presents a real problem in South Africa. The approach taken to addressing energy efficiency here has been different to that in other developing countries. In countries such as Malaysia, Brazil and Morocco, the starting point was with voluntary standards, which were progressively made mandatory. South Africa, on the other hand, borrowed heavily from the Nordic countries in formulating their energy efficiency codes. The authors recognize that local differences require non-uniform specification for performance and their study indicates that local authorities have had a good degree of success in enforcing energy efficiency codes for buildings. Again this indicates that local control has a key role to play in reaching the sustainable building goal. The voluntary versus mandatory debate, as well as the central as opposed to local control question, both surface in this chapter. The authors also acknowledge that in order for the green building agenda to be driven forward in South Africa it is necessary to demonstrate the benefits that increased energy efficiency can yield. Their work identifies a research gap in this respect in South Africa as the necessary data on energy efficiency are not currently available.

Australia and Asia

The first chapter in this final section looks at some of the challenges faced in planning on the urban periphery of Sydney, New South Wales. Andrew H. Kelly is a senior academic, qualified in both law and planning, at the University of Wollongong, New South Wales and Stuart J. Little is a senior environmental planner, currently the Rural Lands Program Coordinator with the Sydney Catchment Authority. It is not only buildings that interact with their environment, but also their surrounds. In Australia, the climate and culture tend to result in homes extending into their gardens, which have been referred to as 'outside rooms' (Hall, 2007). This chapter considers some of the challenges faced in the design of sustainable front and back yards on the outskirts of Sydney. Three elements of modern planning are examined in this respect; enhancing and protecting amenity, conserving biodiversity and minimizing threats from bushfire. There are some interesting comparisons to be drawn between the planning regimes of the UK, the US and Australia. Planning law in Australia very much followed the UK model until the 1940s, but then diverged by maintaining a strict zoning approach. A common feature of all three planning regimes, however, is the evolution of the acknowledge-ment of environmental issues in planning. This is clearly charted in Leiter's chapter and a similar process occurred in both UK and Australian planning law, albeit at a slightly later point. Leiter also acknowledges in his chapter the challenge posed by wildfire to planning in the US. Kelly and Little, in their chapter, focus upon the role of local government in addressing these chal-lenges. It is interesting to note that, although local government clearly has a key role to play in planning, this third tier of government has no formal recog-nition in the Australian Constitution. There are obvious tensions between what is safe, what is visually pleasing and what is environmentally acceptable in theplanning of urban front and back yards and the authors conclude that it is vital that more attention is given to the design of a building's surrounds in the planning process.

The chapter from China is authored by Michael Guan Rui, who has recently completed his doctorate at the Architecture Department at the University of Hong Kong. He is Associate Director of Ho & Partners Architects, Engineers & Development Consultants Ltd. Because of the size and climatic diversity of China, it is divided into five climatic zones and building regulations are designed accordingly for each zone, as are environmental and energy conservation issues. A system of building control has been in place in China since the 1960s and standards for green buildings were introduced in 2006. There is now a comprehensive, if somewhat complex, system in place, with three levels of control over all aspects of buildings. However, only the basic level of control is compulsory. Like many of the other contributors, Guan Rui stresses the need for localized control. This is particularly crucial in a country of the size and diversity of China. Although he sees a role for national central measures, Guan Rui argues that this should be restricted to overarching fundamental law with the bulk of control lying at a local level. The author identifies a move over recent years in China to a much more holistic approach to building sustainability. Previously the focus had solely been upon the artificial built environment and energy efficiency measures, but increasingly standards also address natural environmental load and support capacity. However, Guan Rui is of the view that in order to attain the required standards across all aspects of building, more compulsory provision is required alongside incentives for reaching higher levels. Guan Rui, like Talukhaba *et al*, recognizes the value of indigenous design and techniques. Many of the buildings heralded as 'green' in China are landmark signature constructions employing high levels of technology. Designs and techniques in vernacular architecture are, in the author's view, overlooked. More emphasis, he believes, should be placed upon 'normal green buildings'.

As in China, green building control in Taiwan is fairly comprehensive and well developed. The joint authors of the chapter from Taiwan are Jui-Ling Chen, the Acting Director General of the Architecture and Building Research Institute of the Ministry of the Interior of Taiwan, and Chiung-yu Chiu, who is the Deputy Secretary General of the Taiwan Green Building Council. Green building design was voluntary until 2001, when it became mandatory for all publicly owned buildings over a given value. Today there are mandatory standards in place in Taiwan for both publicly and privately owned new build. Chen and Chiu chart the evolution of green building in Taiwan over three phases, technology research, policy promotion and regulation implementation. In the technology research phase, a green building assessment system with nine indicators was developed to meet the unique climatic conditions of Taiwan. Importantly, alongside this a material evaluation system was also developed, which also addresses environmental impacts (an issue raised in the chapter by McCuen and Fithan). During the policy promotion phase, subsidies and incentives were introduced and in the regulation implementation phase standards were incrementally placed upon a mandatory footing. The authors acknowledge that existing building stock in Taiwan remains a challenge and this seems to be a fairly common picture, which is also recognized in the contributions from the UK and the Netherlands. They also note that there is still work to do in Taiwan in relation to the private sector.

Singapore is a nation that faces particular challenges with energy and water supplies. It relies heavily upon imported oil and water and thus there is a strong imperative to increase energy and water efficiency. The contribution from Singapore comes from a senior academic from the National University of Singapore. Asanga Gunawansa is also an Attorney-at-Law of the Supreme Court of Sri Lanka. This final chapter provides detail of Singapore's regulatory and voluntary provisions for green building standards. As with many other jurisdictions, energy efficiency standards are set in Singapore through building control provisions. In addition, there are two major green building labelling schemes in place to encourage good practice beyond the standards set in the building codes. The Green Mark scheme has now been put on a mandatory footing. The other labelling scheme, the Energy Smart scheme, remains voluntary, although a certain level of attainment under this scheme is required to meet some of the standards for the Green Mark scheme. The chapter from Singapore also touches on an element of green building design not previously considered in detail in the book. South East Asia, where Singapore lies, is one of the most vulnerable regions to the effects of climate change. Gunawansa argues that, in addition to prevention, mitigation and adaptation to the effects of global warming are essential. Green building, he contends, should incorporate the capacity to adapt to a changing climate, for example, by introducing new building techniques to withstand adverse weather conditions and choosing the sites for new construction with care.

Common themes

This book presents a snapshot of the law relating to green buildings in a limited number of jurisdictions across the world. It is by no means comprehensive and because of the broad approach taken to green buildings, a wide variety of different types of legal and voluntary provisions is contained within its chapters. It is, therefore, not sensible to attempt to make direct comparisons either between countries or the different laws and policies. However, there are quite a number of common themes that emerge from the contributions and these are worthy of further brief examination.

(A) 'Green' or 'sustainable' buildings

In this introduction and many of the chapters of the book, the terms 'green building' and 'sustainable building' are used interchangeably. Theron and Dowden, in Chapter 6, ask the question; 'Is sustainability green?' and this is, indeed, an important question and an appropriate starting point to this discussion. In fact 'green' and 'sustainable' are two very different concepts, the latter being drawn from the principle of sustainable development as encapsulated in the Brundtland Report, 1987: 'Sustainable development that meets the needs of the present without compromising the ability of future generations to meet their needs'.

As Theron and Dowden note, the Bruntland Commission gave equal weight in the principle to environmental, social and economic factors. This is not generally the case when considering 'green' or 'sustainable' buildings. Normally speaking it is just the environmental impact of the building that is considered. Gibberd, in his chapter, actually calls for a wider interpretation of 'sustainable' in relation to building standards, one that encompasses the health and well-being of individuals and thus extends into the 'social' sphere of the principle. However, the current understanding is that in the context of building performance 'green' and 'sustainable' are synonymous and relate to the environmental impact of the building. Theron and Dowden provide a good workable definition:

> The practice of 'green building' is seen as the practice of creating structures and using processes that are environmentally responsible and resource efficient throughout a building's life cycle from siting to design, construction, operation, maintenance, renovation and deconstruction.

Having limited the definition to the environmental impacts of buildings, there is still debate as to how widely those impacts should be drawn. The definition above is a broad one and this mirrors the approach taken by this research group and presented within the chapters of the book. All aspects of the interaction between buildings and their environments are considered. Some authors, however, pick up upon a tendency to restrict the remit of green buildings to the artificial built environment and energy efficiency measures rather than address such issues as natural environmental load and support capacity. Guan Riu identifies this in China, although he notes that increasingly a more holistic approach is being adopted there. The lack of holistic approach is also, very much, the central theme of the chapter by McCuen and Fithian. They identify tensions in the United States between the aspiration of 'green building' and the underlying environmental consequences of certain materials, processes and practices, as well as the occupancy phase of a building's life. For a building to be 'green' in accordance with the definition above, all these elements need to be considered. Gunawansa, in the final chapter of the book, suggests that the scope of the definition should be drawn still wider to include the capacity of a building to adapt to climate change.

It is not only drawing the boundaries of the definition that is difficult; measuring the 'sustainability' or 'greenness' of a building presents a fundamental challenge. Murphy picks up on this in Chapter 3, noting that after nearly a decade, the environmental chapter of the Netherlands National Building Decree still remains empty as debate continues on how the sustainability of buildings should be measured. Assessment systems are in place across a number of jurisdictions. See, for example, the Code for Sustainable Homes in the UK and the EEWH Green Building Assessment System of Taiwan. These assessment and rating systems exhibit many common features, but there is as yet no overarching global system of measurement. The Building Research Establishment Environmental Assessment Method (BREEAM) established in the UK has been adopted, to some degree, internationally for non-domestic buildings, but it is almost certainly unrealistic to hope for a comprehensive agreement on a worldwide standard, much as it would simplify matters. Theron and Dowden are, however, optimistic that as the market evolves and new metrics and regulation emerge, so might some consensus on evaluation of the 'greenness' or 'sustainability' of a building.

(B) Voluntary or mandatory standards

A theme that recurs time and again through the book surrounds the relative effectiveness of voluntary as opposed to mandatory performance standards. Talukhaba *et al* note that South Africa would have done well to follow other developing nations in starting with voluntary measures and progressing to those of a mandatory nature. Taiwan has gradually progressed to more mandatory controls, and in China the lower levels of control are mandatory whilst the more advanced levels are not. Guan Riu, however, advocates a transition to more compulsory provisions. In the UK, voluntary measures sit alongside regulation and this has, arguably, proved to be very successful in driving up standards. As can be seen in my chapter, in the UK, performance to a required level in voluntary codes has been used at the planning stage to attain a higher level of performance to that dictated by law and, as the law has since been adjusted to reflect that higher standard, there is strong evidence to suggest that the voluntary measures have been instrumental in driving standards upward. In the Netherlands also, voluntary agreements are being used alongside legal provisions. Although it is too early as yet to judge the performance of the covenants recently introduced to achieve energy performance improvement in buildings, Murphy notes that they show promise and could serve as a useful test-bed for subsequent legislation.

Much will probably depend upon the nature and the level of development of the particular jurisdiction in question. The Netherlands, for example, as Murphy explains, is a forerunner in environmental protection terms and has a tradition of maintaining many of its environmental policies outside the statute books, respecting a common Dutch approach to deliberation and self-regulation. Developing countries, on the other hand, may need to rely upon strong legislative provision with the threat of sanctions to achieve the required result. There are, in fact, two different questions to consider here. First, whether, in introducing standards for sustainable buildings, it is better to start with the voluntary approach and work towards binding standards and second, whether voluntary measures have a role to play alongside standards contained in regulation. In terms of the first question, with the strong imperative of global warming, it is tempting to advocate a launch directly into the regulatory phase. However, there does seem to be a message to be drawn from amongst the contributions to this book that starting with a voluntary approach may be preferable. Nonetheless, it would also appear that a fairly swift transition to regulation is necessary, and incentives and encouragement are also required. On the second question, the experience from the UK and the Netherlands would seem to indicate that voluntary standards and regulation can work well together; the voluntary code ratings encourage innovation and provide the drive to strive for better standards and develop new and affordable technologies. Also, mechanisms such as covenants can test requirements that can subsequently be transferred to legislation.

(C) Local or central control

An interesting picture emerges from the planning systems of three juris-dictions from chapters in the book. In the UK there has been something of a swing back and forth between central and local control in the planning regime over the years. The 1960s and 1970s saw a time when decisions were very much locally based. In the 1980s planning control shifted to the centre. This was a time dominated by the free market and the removal of restrictions upon developers. Control remained with the centre, albeit regional strategies were introduced in 2004 and measures put in place to encourage active engagement with stakeholders and the public. The Planning Act 2008 further centralized major planning decisions, by streamlining the process for major infrastructure projects. However, new legislation proposed by the current government promises to reverse this trend and return planning policy and decisions to local bodies.

Leiter identifies three periods in the United States. The first of these, from 1970 to around 1985, was dominated by federal and state regulations that were implemented directly. In the second period, from 1985 to 2003, the federal and state regulations were addressed at a regional level. Environmental systems plans adopted and implemented through a combination of federal, state, regional and local agencies. The third period, from 2003 to date, has seen a further movement in the US towards integrated regional and local planning, now often directed towards long-term sustainability goals.

From the 1950s, in New South Wales (NSW), Australia, it was the state government that drove metropolitan planning for the Sydney region. The 1980s in NSW saw the emergence of environmental considerations in planning and, at the same time, local communities gained a greater input into plan preparation and implementation. As in the UK and US, with a recog-nition that many issues apply across administrative borders, a regional over-view of planning was also adopted. New South Wales currently has two levels of planning instruments, one based at state /regional level and the other comprising local-level plans. However, the template for local planning demands a high level of conformity and it has been argued that this erodes creativity in local plan making.

A similar picture has unfolded in each of these jurisdictions in that, over time, it appears to have become recognized that there are important national, regional and local elements to sound and sustainable planning decisions. Interestingly, however, in the UK, although there is currently a strong move towards local decision making on planning matters, the regional element appears to have been discarded. As we see above, there are also doubts about the degree of actual local input in NSW. It is only in the US, specifically in the San Diego region that, in Leiter's view, 'federal and state laws have created a framework in which regional and local planning agencies have been able to address community goals and concerns that are unique to this region'.

Another issue relating to central as opposed to local control arises with building standards. There is some variance in practice and opinion amongst the contributions to this book on this question. In the UK, it would seem that the preference is for centrally set uniform standards that are locally enforced. As noted above, the role of the planning system in recognizing and catering for local differences in the UK is under threat. In South Africa, local differences are catered for and Talukhaba suggests that in that country local enforcement has been relatively successful. In China, because of its size and diversity, different standards are applied to different zones, and again, Guan Rui argues that local control has met with some success. In contrast Taiwan appears to have taken a more centralized approach, although some successful local initiatives are acknowledged. In the attempt to meet ambitious global and local targets for reduced carbon dioxide emissions, it is understandable that a single centrally imposed set of standards for buildings is attractive. However, this can ignore important variations in the local environment and stifle innovative local initiatives.

(D) Challenges for the future

Whether the approach taken is a central or a local one, whether voluntary or mandatory, or a combination of all of these, what emerges from these chapters is the fact that there is still much work to be done. In particular, it is acknowledged on multiple occasions that the private sector and existing building stock remain stubborn problems in terms of the improvement of standards of building performance. Education is a key challenge. If the benefits of green building are successfully communicated and these are transferred into increases in property value, then this will provide an important driver to increase standards. Acquiring the data to support the argument for green buildings is, in some countries, a problem and this is an area for possible further research. The aim of this research project was to produce a publication containing a selection of outputs that would allow important lessons to be drawn by nations across the globe at all stages of development. It is hoped that this book will provide some possible directions of travel in order to address the challenges that still exist for green building legislation and practice in its widest sense.

Part I

1 Planning and policies for sustainable development in California and the San Diego region

Robert A. Leiter

1.1 Introduction

Continued rapid urban development in California has placed ever greater strains on the natural environment. In response, conventional methods of urban planning, development regulation and environmental review are being supplemented by new approaches that focus on large-scale natural systems rather than local jurisdictional boundaries or individual projects. These approaches are being applied in the areas of habitat conservation, water resources, air quality, and now climate change. More recently, regional and local governments have begun to use comprehensive plans to address environmental systems' needs in the overall context of *sustainable development*.[1]

This chapter provides an overview of the evolving legal and policy framework in which environmental systems planning and comprehensive planning is being conducted in California, with a focus on planning and policy approaches being taken in the San Diego region. First, the legal framework for considering the natural environment in urban planning, at both the federal and state levels, is reviewed. Next, the evolution in California of approaches to urban planning that led to a greater emphasis on sustainable development is discussed. Finally, a case study of the San Diego region is presented in order to illustrate how this evolution toward sustainability in urban planning and development is taking place in a major metropolitan region in California.

1.2 Background

Scholars who have traced the evolution of environmental planning and regulation in the United States have identified the early 1970s as a major turning point. According to one author, "the first Earth Day on April 22, 1970, brought national attention to the poor environmental quality: 60 percent of America's waterways were not fit for swimming or drinking and many city dwellers choked in smog. The problems were simply too big for cities, metropolitan regions, or even states to handle. Moreover, the private sector had to be included in environmental planning and regulation.

Beginning in 1970, Congress and President Nixon responded with the most sweeping environmental regulation in the history of the United States" (Daniels, 2009: 185). In many instances, environmental actions were taking place at both the federal level and state level nearly at the same time. However, California would often take action that went beyond federal legislation.

1.2.1 Laws protecting environmental resources

The Clean Air Act Extension of 1970,[2] the Water Pollution Control Act Amendments of 1972 (better known as the Clean Water Act),[3] and the Safe Drinking Water Act of 1974[4] created the legal framework under which the U.S. Environmental Protection Agency (U.S. EPA) – created by President Nixon in 1970 – began to implement an aggressive regulatory approach to clean up air and water pollution. At around the same time, the federal Endangered Species Act[5] (ESA) was enacted in 1973 giving authority to federal agencies to regulate activities that could lead to plant and animal species becoming extinct. In California, laws and regulations were enacted during this same period which supplemented federal laws and provided the regulatory framework in which federal policy protecting natural resources could be implemented effectively at the state level.

(A) Air quality

Under the federal Clean Air Act, U.S. EPA was given the authority to set limits on certain air pollutants, including setting limits on how much pollution can be in the air anywhere in the United States. EPA must approve state, tribal, and local agency plans for reducing air pollution. In California, the California Air Resources Board (CARB) was given responsibility for preparing the State Implementation Plan (SIP), which outlines how the state will control air pollution under the Clean Air Act. At the same time, regional air pollution control agencies were directed to prepare and implement regional components of the SIP, and to regulate "stationary" sources of air pollution within their jurisdictional boundaries, such as factories and other businesses that emit large amounts of pollution from one source.

(B) Water quality

In 1972, the federal Clean Water Act established the National Pollutant Discharge Elimination System (NPDES) permit program to regulate the discharge of pollutants from "point sources."[6] In California, attention to water quality through legislation began three years earlier. The Porter-Cologne Water Quality Act[7] first established a regulatory programme to protect water quality and the beneficial uses of state waters. Enacted in 1969, it created and designated the California Water Resources Control Board

(CWRCB) and nine Regional Water Quality Control Boards (RWQCBs) as the principal state agencies responsible for water quality control. Each RWQCB was responsible for preparing and implementing a regional Basin Plan that designated beneficial uses for surface and ground waters; sets narrative and numerical objectives that must be attained or maintained to protect the designated beneficial uses and conform to the state's anti-degradation policy; describes implementation programmes to protect the beneficial uses of all waters in the region; and describes surveillance and monitoring activities to evaluate the effectiveness of the Basin Plan.

(C) Wildlife habitat protection

The federal Endangered Species Act (ESA) of 1973 prohibits any willful taking of threatened or endangered species on all lands within the United States, both public and private. "Taking" is defined as killing, hunting, harming, capturing, or collecting a threatened or endangered species, or destroying its habitats. The U.S. Fish and Wildlife Service (U.S. FWS) was given the primary responsibility for administering the ESA. In 1984, the California Endangered Species Act (CESA) was enacted, giving authority to the California Department of Fish and Game (CDFG) to protect animal and plant species within the state. Since that time, the state and federal agencies have worked in collaboration to enforce the provisions of both laws throughout the state.

1.2.2 Laws requiring local general plans

At the same time that the federal government and state governments were enacting legislation protecting environmental resources, many states began to require that local governments adopt comprehensive plans that incorporated environmental goals and policies. In California, the legislature had first authorized local governments to adopt "master plans" for their communities in 1927. This paralleled efforts across most of the United States to enable planning and zoning. In 1971, the state took a significant step forward by passing a law requiring counties and most cities to bring their zoning ordinances and subdivision procedures into conformance with their "general plans, and requiring that all new development projects and capital improvement projects be consistent with these plans" (Fulton and Shigley, 2005: 105). In addition, the content requirements for local general plans were more clearly defined in the law. Every local general plan was required to contain seven mandatory elements (land use, circulation, housing, conservation, open space, noise, and safety). The legal requirement for general plans to address conservation and open space placed greater attention on environmental resource issues in local government land use planning and regulation.

1.2.3 Laws requiring environmental assessment of plans and projects

Also at this time, federal and state laws were enacted that required that an environmental assessment be conducted on all major development projects and plans, and that mitigation measures for significant environmental impacts be identified and implemented to the extent feasible. At the federal level, the National Environmental Policy Act (NEPA),[8] which was enacted in 1969, required that all federal projects, funding, permits, policies, and actions be screened for environmental effects.

The California Environmental Quality Act (CEQA),[9] enacted in 1970, requires local governments and state agencies to consider the potential environmental effects of a project before making a decision on it. The purposes of CEQA were to disclose the potential impacts of a project, suggest methods to minimize those impacts, and discuss project alternatives, so that decision-makers would have full information upon which to base their decisions.

1.3 The evolution of environmental planning and urban planning in California

The overall approaches to environmental planning and regulation, urban planning and development regulation, and environmental assessment in California have undergone a significant evolution since 1970. This evolution is closely aligned with the evolution of the modern environmental movement at the national level over this 40-year period, which has been documented by a number of planning scholars (Daniels, 2009). In both cases, three distinct "evolutionary periods" can be identified:

- The first period, from 1970 to around 1985, was dominated by federal and state laws and regulations that were implemented directly by regulatory agencies, primarily on a *project-by-project* basis.
- During the second period, from 1985 to around 2003, the implementation of federal and state environmental laws moved toward regional-scale *environmental systems plans*, such as habitat conservation plans, watershed management plans, and regional air quality plans, which were implemented through a combination of federal, state, regional, and local agencies.
- During the third period, from 2003 to the present, while regional-scale environmental systems planning and regulation has continued, there has also been a movement toward *integrated regional and local planning* that coordinates transportation and land use planning strategies with environmental protection strategies, and also looks at how other local and regional systems interact. These integrated plans in many cases are being directed toward long-term *sustainability* goals.

1.3.1 First evolutionary period (1970 to 1985)

(A) Planning and regulation for environmental resource protection

In this first period in California, environmental protection laws were implemented primarily by federal and state agencies along with regional agencies that were charged with enforcing these laws at the regional level:

- *Air quality* – The U.S. EPA worked with CARB and regional air pollution control agencies to implement the federal and state Clean Air Acts, focusing on regulation of mobile sources at the federal and state level, and regulation of stationary sources on a project-by-project basis at the regional level. In 1978, the California Energy Efficiency Standards for Residential and Nonresidential Buildings[10] were established in response to a legislative mandate to reduce California's energy consumption while recognizing that these standards would also lead to reductions in air pollution. The energy standards are implemented through local government building codes, and are updated periodically to allow inclusion of new energy efficiency technologies and methods.
- *Water quality* – The U.S. EPA worked with CWRCB and its nine regional boards to implement the federal Clean Water Act and the state's Porter-Cologne Water Quality Act, with a focus on regulating "point sources" such as sewer plant and industrial discharges on a project-by-project basis.
- *Wildlife habitat protection* – U.S. FWS and CDFG worked together to enforce the federal and state Endangered Species Acts, primarily on a project-by-project basis. One author has observed that during this period, most disputes over impacts of development projects on endangered species involved activities on federally owned land rather than privately owned land; for example, disputes over the spotted owl in Northern California dealt almost exclusively with logging practices on federal land (Fulton and Shigley, 2005: 377).

(B) Local general plans and development regulations

At this time, while local governments in California were required to address the protection of natural resources in their general plans, their general plan policies regarding environmental resource protection were typically not connected directly to implementation of federal or state environmental protection laws. As a result, environmental planning efforts at the community and project level often led to piecemeal actions that were not effective in protecting resources from a systems perspective. In addition, there was little effort to monitor the success of these actions over time.

(C) Environmental assessment

During this period, public agencies were required to assess the environmental impacts of development projects pursuant to NEPA and CEQA, but these project-level analyses were focused on site-specific impacts, and normally did not lead to mitigation requirements that addressed system-wide functionality or long-term viability.

(D) Outcomes

By the mid-1980s, there was growing discontent with the negative impacts of rapid development on natural resources in urbanizing areas of California and elsewhere. Combined with the perceived ineffectiveness of local general plan policies and project-by-project environmental assessment in addressing these concerns, this situation led to calls for more direct intervention by federal and state agencies into local land use and environmental regulation. However, by this time, federal and state funding for implementing environmental regulations was not keeping pace with demands for governmental intervention and oversight. As a result, environmental stakeholder groups began to mount legal challenges against local general plans and land use regulatory decisions for not adequately addressing the cumulative impacts of local development projects on natural systems that went beyond project or community boundaries.

1.3.2 Second evolutionary period (1985 to 2002)

(A) Planning and regulation for environmental resource protection

In response to the problems outlined above, public agencies and private stakeholders began to place a greater emphasis on regional-scale environmental systems planning in the areas of air quality, water quality, and wildlife habitat protection.

(A) AIR QUALITY

Although significant progress was being made in California in improving air quality through the combined efforts of U.S. EPA, CARB, and regional air pollution control agencies, air quality standards were still not being met. One of the legislative responses to this problem at the federal level was contained in the Intermodal Surface Transportation Efficiency Act of 1991[11] (ISTEA), which authorized expanded funding of $150 billion over a six-year period for federal transportation programs. Among the stated purposes of this legislation were:

- to develop new ways of moving people and goods so the Nation can compete in the global marketplace, meet clean air requirements and reduce reliance on foreign oil;

- to provide an improved transportation system for all segments of the population, including elderly, disabled, and other transit-dependent groups; and
- to improve the quality of life in the United States through economic benefits from higher productivity growth, reduced air pollution, and reduced traffic congestion.[12]

Under ISTEA, every metropolitan area with a population of more than 50,000 was required to work through a Metropolitan Planning Organization (MPO) in order to qualify for federal transportation funding, and each MPO was required to produce a 20-year Regional Transportation Plan (RTP) and to update it periodically. As a result, MPOs and regional air pollution control agencies were now required to work together to develop regional strategies to reduce air pollution resulting from auto travel, such as greater investment in public transit and transportation demand management measures.

Similarly, at the state level, in 1997 the California Legislature enacted a law (SB 45[13]) that gave MPOs the authority to programme all long-range state and federal capital investment funds allocated for metropolitan regions in the state—75 percent of all such funds statewide (Barbour and Teitz, 2009: 181). Investments in public transit in California increased significantly after SB 45; in the first four years after the bill's passage, California alone accounted for more than half of all federal funds spent on transit.

(B) WATER QUALITY

In response to continued water quality problems in urban areas, in California CWRCB and its nine regional boards began to place greater attention on "non-point source" water pollution, through the requirement for regional or subregional watershed plans and regulations to manage stormwater runoff caused by urban development. Pursuant to federal and state laws, cities and counties were required to obtain NPDES permits that set general standards for the quality of stormwater runoff. Under these permits, the cities and counties were then responsible for issuing permits to developers for most construction projects. This regulatory structure led to the creation of model regulations that would promote the use of "low impact development" design and construction practices that could be applied consistently by all local governments within a region.

(C) WILDLIFE HABITAT PROTECTION

Also at this time, rapid urban development and diminishing open space in Southern California were leading to an inevitable conflict. "At the center of the conflict was the fate of more than 340,000 acres of coastal sage scrub occupied by the coastal California gnatcatcher, a small songbird whose range extends across San Diego, Orange, Riverside, San Bernardino, and Los

Angeles counties. Environmentalists petitioned state and federal wildlife agencies to designate the gnatcatcher as endangered, against the opposition of the development community" (Cylinder *et al.*, 2004: 3). To address this conflict, the California Legislature passed the Natural Community Conservation Planning Act in 1991[14] (NCCP Act). The NCCP Act provided for a regional planning process focused on protection of biological communities rather than single species. The goal of the act was to conserve species before they became endangered. The law set forth requirements for preparation and adoption of Natural Community Conservation Plans, similar to "habitat conservation plans" (HCPs), which were allowed under the federal ESA, but had rarely been used. These plans were designed to cover a regional geographic area, and allowed for the protection of one or more sensitive plant and animal species within that area, while also allowing less sensitive areas to develop. These plans were typically developed by a local or regional public agency or private property owner, in collaboration with U.S. FWS and CDFG and environmental stakeholder groups. By 2003, 127 habitat conservation plans had been prepared in California.

(B) Local general plans and development regulations

During this same period, a second generation of local general plans, community plans, and specific plans developed by cities and counties in California were placing greater attention on the cumulative impacts of urban development on environmental systems and urban service systems. Many local governments also implemented *performance-based* growth management systems, tied to their general plans, to address the impacts of urban growth on their communities.

(C) Environmental assessment

With the increased focus on environmental systems planning, there were parallel efforts to streamline the environmental assessment process for development projects by preparing "programmatic" environmental documents for local general plans and for environmental systems plans. It was argued that, in this way, the programmatic environmental documents could address the cumulative impacts of development within a given geographic area on environmental systems, and identify programmatic mitigation measures that in many cases would simply require development projects to be consistent with the plans and/or implement specific actions at the project level. This "tiered" approach to preparation of environmental impact reports was incorporated into CEQA in 1993, and has been used with limited success since that time.

(D) Outcomes

During the second evolutionary period from 1985 to around 2002, the role of regional transportation planning agencies in California was expanded significantly to include responsibility for conducting multi-modal regional transportation planning with significant stakeholder input, and to address air quality issues. At the same time, there was greater use of regional environmental systems planning in California and other parts of the United States, particularly in the areas of water quality protection and wildlife habitat protection. However, there were also growing concerns about the long-term financial viability of implementing these plans, which relied on ongoing management, enforcement, and monitoring activities. In addition, there were concerns about negative economic impacts that these programs could be having in regions where they were being implemented, and about the effects of these programs on overall housing affordability and social equity.

1.3.3 Third evolutionary period (2003 to present)

(A) Planning and regulation for environmental resource protection

In response to the continued challenges of addressing environmental resource protection and public facility needs in urbanizing regions, in the late 1990s, regional transportation planning agencies in California began to develop and adopt "regional comprehensive plans" or "regional blueprint plans." Each of these plans was developed independently by an MPO working with its local governments, as well as with state and federal agencies. These plans were designed to create an integrated planning framework that looked at the connections among transportation, land use, housing, economic development, environmental resource protection, and public facilities. In most cases, these plans were developed through a "scenario planning" process in which alternative futures were evaluated, and a preferred planning concept built on principles of sustainability was formulated with extensive stakeholder participation and public input. In 2008 the State Legislature enacted SB 375,[15] which requires each MPO to prepare a "Sustainable Community Strategy" (SCS) in conjunction with any future update to its RTP. The purpose of the SCS is to demonstrate how the greenhouse gas emissions associated with future development in the region may be reduced through land use and transportation strategies. The SCS will contain many of the same components as existing regional blueprint plans, but is also required to demonstrate how the region will meet greenhouse gas emission targets for 2020 and 2035 to be set by CARB. The SCS is also required to address protection of environmentally sensitive lands, and must be coordinated with the "Regional Housing Needs Assessment," which each regional planning agency in California must prepare periodically under state planning law, to provide guidance to local governments on how much land should be planned and zoned for housing in each community.

(B) Local general plans and development regulations

At the same time that regional planning agencies are developing comprehensive plans that address sustainability, the state is encouraging local governments to place a greater emphasis on sustainability and "environmental justice"[16] in their general plans. In response to an amendment to state planning law,[17] the latest edition of General Plan Guidelines published by the Governor's Office of Planning and Research includes an entire chapter on "Sustainable Development and Environmental Justice," and provides specific guidelines for methods that local governments can use to address specific environmental justice issues in their plans (Governor's Office of Planning and Research, 2003: 20–31).

(C) Environmental assessment

During this period, environmental assessments for regional plans and local general plans have received greater attention from the state. For example, the State Attorney General's Office has challenged the adequacy of Environmental Impact Reports[18] (EIRs) for regional and local plans that in its view did not adequately address the impacts of new development on greenhouse gas emissions.

Also, in conjunction with the movement toward using regional blueprint plans to integrate planning policies related to greenhouse gas emissions and other statewide goals, there has been a renewed effort to streamline environmental assessment requirements for individual projects. Specifically, SB 375 contains provisions that allow for reduced environmental assessment reporting for individual development projects where a Sustainable Community Strategy has been approved by the region's MPO and where specified planning criteria are met.

(D) Outcomes

Since 2002, MPOs for the four major metropolitan regions of California (San Francisco Bay metropolitan area, Greater Los Angeles metropolitan area, Sacramento, and San Diego) have developed regional blueprint plans, and MPOs in other parts of the state have begun to prepare similar plans. The fact that SB 375 relies on the use of such plans to address state greenhouse gas reduction requirements suggests that these plans will gain in importance during the coming years. In a recent analysis of the California regional blueprint planning program, the authors observed that "SB 375 establishes a new state framework for metropolitan growth planning in

California that largely follows the blueprint model" (Barbour and Teitz, 2009: 194).

1.4 Case study: planning in the San Diego region

The evolution of federal, state, regional, and local urban planning and environmental planning programs and activities in the San Diego region from 1970 to the present provides an informative case study that can be used to illustrate and explain many of the trends that have been observed in California and the United States as a whole over the past forty years.

1.4.1 Background

The San Diego region (consisting of the County of San Diego and its eighteen cities) contains 4,260 square miles (11,035 square kilometers), with most of the region's population having settled within the coastal plain and adjoining valleys in the western 20 percent of the region. The vast majority of growth in the region has occurred since 1940, the result of a large military presence during World War II and a post-war population boom driven by the aerospace industry through the 1950s and 1960s, with a more diversified economy based on services, research and development, manufacturing, and tourism evolving from the 1970s to the present. Population estimates for San Diego County and California from 1930 to 2010 can be seen in Table 1.1.

It should also be noted that the San Diego region adjoins the Greater Los Angeles metropolitan region (the largest in the United States) to the north, and the Tijuana–Northern Baja California metropolitan region to the south.

1.4.2 First evolutionary period (1970 to 1985)

(A) Environmental resource protection in the region

During the period from 1970 to 1985, the approach taken by the federal and state agencies that were assigned to enforce environmental protection laws in the San Diego region was similar to that taken in other parts of the state, with those agencies implementing laws and regulations primarily on a case-by-case basis.

(B) Local general plans and development regulations

As discussed earlier, state laws governing local general plans were amended in the early 1970s to require the inclusion of several mandatory elements, as well as requiring that all development projects and all capital improvement projects be consistent with local general plans. The county of San Diego and its cities were all responsible for preparing and implementing plans that complied with the new state laws.

Table 1.1 Population estimates

	1930	1940	1950	1960	1970	1980	1990	2000	2010
County	209,659	289,348	556,808	1,033,011	1,357,854	1,861,846	2,498,016	2,813,833	3,224,432
State	5,677,251	6,907,387	10,586,223	15,717,204	19,971,069	23,667,764	29,760,021	33,871,648	38,648,090

Source: Adapted from San Diego Historical Society website (https://www.sandiegohistory.org/links/sandiegopopulation.htm)and California Department of Finance, "Report: California Added 393,000 in 2009; Population Tops 38.6 Million," April 29, 2010.

(C) Environmental assessment

Local governments were also required to comply with provisions of CEQA that called for preparation of EIRs for public or private projects and plans that could have a significant impact on the environment. The state developed administrative guidelines to assist local governments in implementing CEQA. However, as with state planning laws, the primary responsibility for interpreting and complying with CEQA rested with local governments.

1.4.3 Second evolutionary period (1985 to 2002)

In response to significant community concerns about the impacts of growth in the San Diego region and resulting traffic congestion, regional leaders responded with two initiatives:

• In 1988, the San Diego Association of Governments (SANDAG), which serves as the regional transportation planning agency for the San Diego region, obtained voter approval for a half-cent sales tax for transportation (known as TransNet), which provided $3.3 billion in funding over a twenty-year period for regional highway and public transit projects, along with local transportation projects and activities.
• Also in 1988, a voter initiative was passed in San Diego County that led to designation of SANDAG as a "regional growth management" organization, and gave it specific responsibilities to monitor and report on activities being undertaken in the region to address the impacts of growth on public facilities and the environment.

(A) Environmental resource protection in the region

(A) AIR QUALITY

Air quality in the San Diego region improved steadily during this period, largely because of technological improvements that resulted in cars and trucks emitting fewer pollutants per mile traveled, as well as ongoing enforcement of stationary source regulations by the San Diego County Air Pollution Control District. However, as discussed earlier, federal transportation law passed in 1991 required that SANDAG, the designated MPO for the San Diego region, begin to take a broader look at multi-modal transportation options and to evaluate "air quality conformity" in the periodic updates to its RTP. These requirements, combined with SANDAG's new roles in administering the TransNet transportation sales tax program and serving as the regional growth management agency, brought much greater attention to its activities by environmental stakeholders and community leaders in the region.

(B) WATER QUALITY

In 1990, to reduce worsening pollution from urban runoff in the region's watersheds, SDRWQCB began working with the San Diego County, the eighteen cities, and the San Diego Unified Port District (collectively known as "co-permittees") to prepare the region's first water quality management plan, and issued an NPDES Stormwater Permit to enforce implementation of this plan. The permit ordered the co-permittees to collaborate to control waste discharges in stormwater and other urban runoff from the storm sewer systems that drain into the watersheds of the region. This permit was renewed in 2001, and again in 2007, each time evaluating progress being made toward meeting water quality standards in watersheds containing impaired water bodies, and in some cases imposing new requirements on the co-permittees.

(C) WILDLIFE HABITAT PROTECTION

As discussed previously, in 1991 the state enacted the NCCP Act. This legislation allowed the preparation and adoption of two subregional plans in San Diego County. The largest subregional plan, the Multiple Species Conservation Program (MSCP), spans eleven cities and unincorporated portions of central and southwest San Diego County. In 1991, the City of San Diego took the lead in the preparation of the MSCP, which was designed to address the impacts of regional growth on native species and their habitats within a 900-square-mile (2,330-square-kilometer) study area, and to meet the requirements for a Habitat Conservation Plan (HCP) under the ESA and the NCCP Act. Through the preparation of the MSCP, the participating local jurisdictions agreed to designate certain sensitive habitat areas for permanent preservation, in order to protect eighty-five animal and plant species, while allowing development in non-designated areas. Approved in 1997, the plan targets more than 172,000 acres (69,600 hectares) for preservation.

The Multiple Habitat Conservation Program (MHCP) includes seven cities in northern San Diego County. This subregional plan, which was prepared by SANDAG staff in consultation with participating local governments and federal and state wildlife agencies, and approved by the SANDAG Board of Directors in March 2003, provides the guidance for the preservation of a 20,000-acre (8,100-hectare) preserve system and the protection of sixty-one plant and animal species.

San Diego County is continuing to work on MSCP plans for unincorporated areas in the northern and eastern portions of the county. These plans, when completed, will result in a total of more than 300,000 acres (137,600 hectares) of open space being preserved through HCPs in the San Diego region.

(B) Local general plans and development regulations

While systems planning approaches were being undertaken at a regional scale for environmental resources in the San Diego region, local government land use planning efforts were in many cases following the same path. For example, in 1988, the city of Chula Vista and county of San Diego agreed to jointly prepare and adopt a plan for Otay Ranch, a 22,900-acre (9,300-hectare) site located in the southeastern portion of the Chula Vista planning area. In 1993, following several years of planning and stakeholder input, the city and county jointly approved the Otay Ranch General Development Plan/Subregional Plan, which allowed development of 27,000 homes organized into eleven villages, many of which were designed to be transit-oriented and interconnected by a regional transit system. The plan also included an urban center that would contain a variety of office, retail commercial and high density residential uses, and a site for a future public university campus that would be connected to the urban center.

At the same time, the Otay Ranch master plan included an "environmental resource plan" to protect important natural resources through creation of an 11,000-acre (4,500-hectare) open space preserve. Developers were required to dedicate land to this preserve and provide funding to support ongoing habitat management and monitoring, as individual projects were entitled and built. Since the adoption of the plan, more than 8,700 housing units have been built and about 2,300 acres (930 hectares) of open space have been dedicated. In addition, several public parks, schools, and other public facilities and amenities have been constructed to support the community in accordance with a performance-based growth management system developed by the City of Chula Vista and applied both to this project and to other master planned communities developed during the same period.

(C) Environmental assessment

The adoption of regional habitat conservation plans, and master plans for development projects like Otay Ranch, which included environmental mitigation programs, led to greater use of tiered environmental documents. However, environmental assessment continued to be a complex process, and "project level" environmental documents were still required in many cases. In addition, many projects were still being challenged legally by environmental stakeholder groups on the grounds that their EIRs were inadequate.

1.4.4 Third evolutionary period (2003 to present)

In the late 1990s, SANDAG analysts began calling attention to the fact that current local general plans in the region would be unable to accommodate forecasted growth and were inconsistent with the Regional Growth Management Strategy. An intensive, three-year-long debate on regional governance arrangements ensued. In 1999 state legislation was introduced that would have replaced SANDAG with a new regional agency subsuming six existing agencies, to be governed by a directly elected board.[19] Instead, in 2002, the legislature passed Senate Bill 1703,[20] which moderately strengthened SANDAG's authority by transferring planning and project development responsibilities from the region's two transit agencies to it and altering its governance structure. In 2003, Assembly Bill 361[21] was enacted, giving SANDAG the authority and the responsibility to prepare and adopt a "Regional Comprehensive Plan" (RCP) by 2004 that would incorporate public input, use the agency's authority over regional transportation funds to further the goals of the plan, and monitor progress through "realistic measurable standards and criteria" to be included in the plan.

(A) Environmental resource protection in the region

The RCP was designed to build upon the regional transportation plan and the regional-scale environmental systems plans that had been developed during the previous decade, while addressing environmental planning issues in the broader context of an overall comprehensive plan. The RCP effort was spearheaded by SANDAG's Regional Planning Committee, made up of local elected officials representing six subregions in San Diego County and advisory members representing federal, state, and regional public agencies. In addition, SANDAG was assisted in its work by a stakeholder working group, as well as a technical working group made up of local government planning directors. The two-year planning process also included extensive public outreach.

The final draft plan was organized in a manner similar to local general plans, containing a long-range vision for the region, and including elements addressing transportation, land use/urban form,[22] housing, economic development, healthy environment (including air quality, water quality, habitat conservation, and shoreline preservation), regional public facilities (including energy, water supply, and waste management), and interregional and binational issues. The RCP also included a five-year action plan of strategic initiatives to implement the goals and policies contained in the plan, as well as a performance monitoring program that designated a set of regional performance indicators to be monitored on an annual basis. The RCP was unanimously adopted by the SANDAG board of directors in July 2004, and the agency began immediately to implement certain key strategic initiatives contained in the RCP, three of which are discussed below.

(A) TRANSNET ENVIRONMENTAL MITIGATION PROGRAM

In 2003, SANDAG began preparation of a November 2004 ballot measure and expenditure plan to extend the San Diego region's existing TransNet half-cent sales tax for transportation. During the formulation of the expenditure plan, environmental stakeholder groups raised the concern that a transportation funding measure that did not address funding needs for habitat protection could lead to future problems in obtaining regional funding to fully implement the region's recently approved HCPs. As a result, the SANDAG board of directors included policies in the RCP supporting the concept of targeting transportation sales tax revenues to acquire habitat mitigation land for new transportation projects.

SANDAG staff recommended an approach that would allow for mitigation for the environmental impacts of transportation projects through acquisition, management, and monitoring of open space areas that were included in the adopted HCPs. The staff developed mitigation cost estimates for all of the regional projects contained in the Regional Transportation Plan ($450 million) and incorporated those costs into the overall expenditure plan. In addition, staff estimated the economic benefits associated with obtaining project coverage through the HCPs, and recommended including an additional $150 million in the expenditure plan, reflecting the economic benefit of HCP coverage for these projects. A similar methodology was applied to local transportation projects, resulting in an additional $200 million being included, for a total budget of $850 million for the Environmental Mitigation Program, out of a total transportation expenditure plan of $14.4 billion. The TransNet extension was approved in the November 2004 election with 67 percent of the vote, slightly greater than the two-thirds majority needed for approval. The measure received strong support from the environmental community, which was viewed by many as a significant factor in its passage.

(B) SMART GROWTH CONCEPT MAP

The RCP also called for the development of a "Smart Growth Concept Map" (SGCP) that would illustrate the location of existing, planned, and potential locations for smart growth in the San Diego region. Following adoption of the RCP, "smart growth opportunity areas" were identified by SANDAG staff and local government planning directors, using seven "smart growth place types" that were defined in the RCP, and the regional transit network plan contained in the RTP. The SGCP also delineated the open space preserve areas included in the region's adopted HCPs. The SGCP, which was accepted by the SANDAG board of directors in 2006, identified nearly two hundred locations where smart growth development could occur in the San Diego region, 40 percent of which were already allowed by local general plans. The SGCP has led to continued collaboration within the region regarding the different ways in which transit-oriented development can be accommodated in existing communities, and is now being used to help prioritize transportation investments as the RTP is updated periodically. It is also used in the allocation of regional funding for local infrastructure improvements through the TransNet Smart Growth Incentive Program and other related grant programs.

(C) REGIONAL QUALITY OF LIFE FUNDING INITIATIVE

As discussed earlier, a major focus of the region over the past several years has been on obtaining adequate long-term funding to support its habitat conservation plans, water quality management plans, and other environmental initiatives. The RCP included a financial analysis of these regional planning programs, which concluded that even with the TransNet Environmental Mitigation Program, there was inadequate funding to support long-term implementation of these plans, and recommended further evaluation of regional funding sources to support these activities. Since that time, SANDAG has been working toward developing a "Regional Quality of Life Funding Initiative" that would secure long-term funding for habitat conservation, water quality management, shoreline preservation, and public transit operations. In 2008, state legislation was passed[23] that gave SANDAG the authority to enact such a funding measure (which could include an additional sales tax for these purposes) and SANDAG is currently working toward a possible ballot measure in 2012. At the same time, SANDAG and other regional and local entities have sought and obtained funds from other external sources to continue environmental program implementation, in areas such as energy efficiency programs, shoreline preservation programs, and water quality improvement programs and activities.

(B) Local general plans and development regulations

While SANDAG was preparing its Regional Comprehensive Plan, the cities of San Diego and Chula Vista, and the county of San Diego, were updating their general plans. In each case, the general plan update has placed a greater emphasis on long-term sustainability than was reflected in the previously adopted plan.

For example, the strategic framework for the City of San Diego's updated General Plan is called the "City of Villages" strategy.[24] This strategy "focuses growth into mixed-use activity centers that are pedestrian-friendly districts linked to an improved regional transit system. . . . The strategy draws upon the character and strengths of San Diego's natural environment, neighborhoods, commercial centers, institutions, and employment centers. The strategy is designed to sustain the long-term economic, environmental, and social health of the City and its many communities. It recognizes the value of San Diego's distinctive neighborhoods and open spaces that together form the City as a whole." (City of San Diego, 2008: SF-3).

The Conservation Element of the city's updated General Plan reinforces the plan's focus on sustainability. It states that "the City is implementing sustainable development policies that will reduce its environmental footprint, including: conserving resources, following sustainable building practices, reducing greenhouse gas emissions, and encouraging clean technologies. In sustainable development practices, economic growth is closely tied with environmental, "clean," or "green" technologies and industries. San Diego is well positioned to become a leader in clean technology industries due to its highly qualified workforce, world-class universities and research institutions, and established high technology industries. Clean technology industries demonstrate that environmental protection and economic competitiveness goals are aligned and mutually beneficial" (City of San Diego, 2008: SF-24).

(C) Environmental assessment

While regional plans and local general plans have been moving toward better integration and greater attention to long-term sustainability, certain aspects of the legal requirements for environmental assessments of these plans have remained unclear, and public agencies in the San Diego region and elsewhere in the state have continued to sustain legal challenges to their environmental documents.

1.5 Conclusion

Over the past forty years there has been an evolution in the laws and policies governing environmental resource protection in the United States and in California, in which three distinct evolutionary periods have been identified. Through this evolution, environmental and overall sustainability considerations have become more integrated in planning and decision-making. In looking at the progression of environmental resource planning and regulation, local government general plans and development regulations, and environmental assessment practices in the San Diego region during this era, we can see clearly how the federal and state laws have created a framework in which regional and local planning agencies have been able to address community goals and concerns that are unique to this region. At the same time, during this era we have seen how the San Diego region has itself evolved to place a greater importance in its planning on *sustainability* at a variety of scales.

As the first region in California to take on the requirements of SB 375, the state's ambitious law addressing greenhouse gas emissions from the transportation sector, the capacity of the San Diego region to look beyond local and regional interests and consider the big picture of global climate change will certainly be tested. In addition, the region faces many other environmental challenges, including the increasing frequency of wildfires in open space areas adjacent to urban development, and recurring shortages of water to serve the region's growing population. However, this region has created a planning framework, grounded in regional and local collaboration, which should serve it well as it faces these and other challenges.

Notes

1 The 1987 report *Our Common Future* from the United Nations World Commission on Environment and Development first defined the concept of *sustainable development*, as "development that meets the needs of the present generation without compromising the ability of future generations to meet their own needs." The concept also involves seeking a balance of environmental, economic, and social equity values ("the three Es"). Source: Berke *et al.*, 2006: 10–11.
2 Clean Air Act Extension of 1970, Pub. L. No. 91-604, 84 Stat. 1676 (1970).
3 Federal Water Pollution Control Amendments of 1972, Pub. L. No. 92-500, 86 Stat. 816 (1972).
4 Safe Drinking Water Act of 1974, Pub. L. No. 93-523, 88 Stat. 1660 (1974).
5 Endangered Species Act of 1973, Pub. L. No. 93-205, 87 Stat. 884 (1973).
6 "Point source" pollutants are those that originate from an identifiable source or "point" of waste release, such as municipal sewage treatment plant outfalls and stormwater conveyance system outfalls.
7 California Water Code Section 13000 *et seq.*
8 National Environmental Policy Act of 1969, Pub. L. No. 91-190, 83 Stat. 852 (1970).
9 California Public Resources Code Section 21000 *et seq.*
10 California Code of Regulations, Title 24, Part 6.
11 Intermodal Surface Transportation Efficiency Act of 1991 Pub. L. No. 102-240, 105 Stat. 1914 (1991).
12 Pub. L. No. 102-240, 105 Stat. 1914 (1991), Sec. 3.
13 California Senate Bill No. 45, Amending Sections of the Government Code, Public Utilities Code, and Streets and Highways Code, 1997.
14 California Fish and Game Code Section 2800–2835.
15 California Senate Bill No. 375, Amending Sections of the Government Code and Public Resources Code, 2008.
16 *Environmental justice* is defined in California state planning law as "the fair treatment of people of all races, cultures, and incomes with respect to the development, adoption, implementation and enforcement of environmental laws, regulations, and policies" (Government Code Section 65040.12 (e)).
17 California Assembly Bill No. 1553, Amending Sections of the Government Code, 2001.
18 *Environmental Impact Reports* are environmental assessments for public or private projects or plans that could have a significant impact on the environment under the California Environmental Quality Act.
19 Members of the SANDAG board of directors are elected officials from San Diego County and each of the eighteen cities, and are appointed by their governing bodies.
20 California Senate Bill No. 1703, Amending Sections of the Government Code and Public Utilities Code, 2002.
21 California Assembly Bill No. 361, Amending Sections of the Public Utilities Code, 2003.
22 Recognizing that SANDAG did not have regulatory authority over land use, the RCP included recommended ways in which SANDAG could encourage compatible land uses through incentives and collaboration with local governments.
23 California Senate Bill No. 1685, Amending Sections of the Public Utilities Code, 2008.
24 A "village" is defined in the City of San Diego General Plan as "the mixed-use heart of a community where residential, commercial, employment, and civic uses are all present and integrated. Each village will be unique to the community in

which it is located. All villages will be pedestrian-friendly and characterized by inviting, accessible and attractive streets and public spaces . . . . Individual villages will offer a variety of housing types affordable for people with different incomes and needs. Over time, villages will connect to each other via an expanded regional transit system." Source: *City of San Diego General Plan,* 2008: SF-3.

References

Barbour, E. and Teitz, M. (2009) "Blueprint Planning in California: An Experiment in Regional Planning for Sustainable Development," in Mazmanian, D. and Kraft, M., *Toward Sustainable Communities: Transition and Transformations in Environmental Policy.* Cambridge MA: MIT Press.

Berke, P., Godschalk, D. and Kaiser, E. (2006) *Urban Land Use Planning, Fifth Edition.* Urbana IL: University of Illinois Press.

City of San Diego (2008) *General Plan,* San Diego CA.

Cylinder, P., Bogdan, K., and Zippin, D. (2004) *Understanding the Habitat Conservation Planning Process in California.* Sacramento, CA: Institute for Local Self Government.

Daniels, T. (2009) "A Trail Across Time: American Environmental Planning from City Beautiful to Sustainability," *Journal of the American Planning Association,* Vol.75, No. 2, pp. 178–192.

Daniels, T. and Daniels, K. (2003) *The Environmental Planning Handbook for Sustainable Communities and Regions.* Chicago IL: Planners Press.

Fulton, W. and Shigley, P. (2005) *Guide to California Planning, Third Edition.* Point Arena CA: Solano Press Books.

Governor's Office of Planning and Research (2003) *State of California General Plan Guidelines.*

2 Potential harmful environmental impacts as a consequence of material and system specifications, installation, and operations in current U.S. green building practices

Tamera L. McCuen and Lee A. Fithian

2.1 Introduction

Current green building practice suffers from disconnects between the owner/occupier desires and perceptions; the actual means and methods used to construct a building; and the environmental impacts during use. While U.S. market transformation toward green building has occurred utilizing a dollar-value-installed equation, life-cycle effects relating to raw material production (cradle-to-use phase), construction means, and end-use energy savings have largely been ignored. This chapter focuses on the discussion on life-cycle emissions and green buildings from three perspectives that include the primary material production, construction means and methods, and occupied phases of a building's life-cycle. The three perspectives are: (1) emissions/environmental harm from material manufacturing for the construction phase, (2) emissions/environmental harm from equipment/ processes during the construction phase, (3) emission/environmental harm from the completed building (post construction).

Issues in these areas include the characteristics of some of the primary raw materials used to produce basic construction materials and the misconceptions regarding their environmental impacts; concerns about the harmful properties of certain manufactured materials/systems; and concerns about emissions by installed systems over their warranted life-cycle and use with particular regard to fundamental energy production. Governance in these areas is often incomplete and/or inconsistent with the intent of green building. The governing bodies that are in charge of industry standards have membership drawn largely from vested industry participants. These bodies tend to utilize consensus-based voting procedures that leave the standards' development subject to conflicts of interest. Furthermore, discussions regarding energy efficiencies and the focus on greenhouse gas emissions forgo more serious implications of point-source emissions of toxic material.

This chapter is organized in sections and is written with the intent of providing an overview and brief discussion about select materials specifications, system specifications, construction processes, and building operations. The chapter exposes some of the potentially harmful environmental impacts and hopes to initiate more dialogue resulting in speedy resolutions that are devoid of harmful environmental impacts. Ultimately these resolutions will inform and direct future environmental legislation.

2.2 Background

Environmental activism in the United States began early in the nineteenth century and has grown as there has been a convergence of interest in the conservation of natural resources and an interest in public health and quality of life. Increased interest in environmentalism began as a response to the negative environmental impacts of the industrial revolution along with the modern chemical revolution. During the industrial revolution nature was objectified and viewed as an agricultural and economic commodity in which land itself was devoid of its association with nature; it was thought of as separate from nature – only as property (Keeler and Burke, 2009). The technology developed during the industrial revolution provided landowners with multiple means to optimize the land use for profit. Unfortunately this was more often than not done at any cost to the environment with extremely harmful consequences.

Following the industrial revolution was the modern chemical revolution. In 1962 the book *Silent Spring*, by a young biologist named Rachel Carson, exposed the impact of the modern chemical revolution on the biosphere, food chain, water cycle, and ultimately humans (Keeler and Burke, 2009). Carson's book exposed common practices for storage, disposal, and transport in the chemical industry. "Her book prompted discussion and controversy, which eventually gave rise to the establishment of governmental oversight agencies, such as the U.S. Environmental Protection Agency (EPA), in spite of the chemical industry's efforts to discredit and vilify Carson." (Keeler and Burke, 2009)

The EPA was established in July 1970 in response to the growing public demand for cleaner water, air, and land. The federal government realized that it was not structured to make a coordinated attack on the pollutants that harm human health and degrade the environment. As a result the EPA was assigned the daunting task of repairing the damage already done to the natural environment and establishing new criteria to guide Americans in making a cleaner environment a reality.[1]

Environmental legislation passed by legislators and politicians since the 1970s has increased significantly in response to public outrage about the harmful impacts on the environment from the industrial revolution and modern chemical revolution. Examples of such legislation include the Clean Air Act of 1970 (including revisions in 1977, 1981, and 1990); Clean Water Act (Federal Water Pollution Control Amendments of 1972); fuel-efficiency standards for automobiles; Resource Conservation and Recovery Act (1976, and subsequent Federal Hazardous and Solid Waste Amendments of 1984); Safe Drinking Water Act of 1974 (and subsequent amendments in 1986 and 1996); Pollution Prevention Act of 1990; Energy Policy Acts of 1992 and 2005; and regulations on pollution emission controls. This list does not fully encompass the plethora of acts and regulations enacted during the last forty years. Legislation is typically passed for a particular component or problem in the system, rather than addressing the entire system. This lack of a systems approach extends to the built environment and is evident in standards, codes, and regulations set forth by governing agencies that oversee the building design and construction industry.

In 1994 Task Group 16 of Conseil International du Bâtiment (CIB) at the First International Conference on Sustainable Construction in Tampa, Florida, formally defined the concept and articulated the principles of sustainable construction (Kibert, 2005). According to the CIB, the Seven Principles of Sustainable Construction are:

1   Reduce resource consumption
2   Reuse resources
3   Use recyclable resources
4   Protect nature
5   Eliminate toxics
6   Apply life-cycle costing
7   Focus on quality

It is the goal of this chapter to present current practices in the U.S. green building industry and initiate dialogue as a stimulus for actions to resolve inconsistencies evident between the intent of green building and U.S. standards, codes, regulations, and environmental legislation.

2.3 Manufacturing processes of materials used in green building

Green building materials are currently evaluated on a life-cycle basis. Certain materials are touted for their recycled content or least impact or simply accepted for their systemic integration into the products used within the building industry. The discussion on the four materials cited below will focus on those aspects of their life cycles that have significant impact either through human health factors or environmental consequences.

2.3.1 Aluminum

The green building industry and the recycling movement in general in the United States hails aluminum as the recycled content poster child. While recycling is a major consideration in continued aluminum use, less than half of all the aluminum currently produced to meet demand originates from recycled raw materials (USGS, 2009). Aluminum can be recycled over and over again without loss of properties, but current and projected usage will still require the mining of bauxite and the conversion to aluminum through the Bayer process.

Aluminum is the third most abundant element in nature, comprising 8 percent of the earth's crust. The ore from which aluminum is produced is bauxite. More than 130 million tons of bauxite are mined each year, the major deposits being in the tropics and sub-tropics. Bauxite is currently being extracted in Australia, Central and South America (Jamaica, Brazil, Surinam, Venezuela, and Guyana), Africa (Guinea), Asia (India, China), the Commonwealth of Independent States, and parts of Europe (Greece and Hungary). In many of these regions bauxite is the only valuable natural resource.

Alumina, the raw material for primary aluminum production, is extracted from bauxite. Bauxite is processed into pure aluminum oxide (alumina) before it is converted to aluminum by electrolysis. In this process, called the Bayer chemical process, the aluminum oxide is released from the other substances in bauxite in a caustic soda solution, which is filtered to remove insoluble particles. The aluminum hydroxide is then precipitated from the soda solution, washed, and dried, while the soda solution is recycled. After calcination, the end-product, aluminum oxide (Al_2O_3), is a fine-grained white powder. Four tons of bauxite are required to produce two tons of alumina, which in turn produces one ton of aluminum at the primary smelter. In 2003, 59 million tons of alumina were produced world-wide.[2]

Jamaica is the third largest producer of bauxite ore in the world and fourth in the production of alumina. Jamaican bauxite and alumina accounted for about 75 percent of total exports with the United States being the major market.[3] Nearly all bauxite consumed in the United States in 2009 was imported, with 31 percent coming from Jamaica, 22 percent from Guinea, 19 percent from Brazil, 12 percent from Guyana, and 16 percent from various other sources (USGS, 2009a).

The major environmental problem caused by the industry is the disposal of the tailings, which form an alkaline red mud that in the past was stored in dammed valleys and spent ore mines. In Jamaica, these "red mud lakes" resulted in the percolation of caustic residues (sodium) into the underground aquifers in local areas. These sites have never been remediated. Current methods are to build clay sealed ponds that are designed to hold 5–7 years of mud storage; however, these ponds never "dry out" and are basically abandoned. Recent readings obtained from domestic water wells in the vicinity of Jamaican alumina refineries have indicated elevated sodium and Ph readings.

2.3.2 PVC

Primary concerns relating to PVC are introduced during the production and disposal phases. The continued use of PVC despite social equity issues and environmental contamination is typically weighed against life-cycle costs and the almost universal use of PVC within the construction industry. Products typically associated with PVC include conduit, valves, connectors, electrical shielding, window frames, resilient flooring, drain/waste/vent piping, and siding.

In 2003, the U.S. Green Building Council assembled a technical advisory group to discuss the life-cycle impacts of PVC and to promote the discussion as to a possible incentive point in the LEED (Leadership in Energy and Environmental Design) system that would promote the avoidance of PVC in the built environment. The task group investigated "whether for those applications the available evidence indicates that PVC-based materials were consistently among the worst of the materials studied in terms of environmental and health impacts" (Altschuler *et al.*, 2007). This task force assembled over 2,500 studies relating to the life cycle and toxic effects of PVC.[4]

The study went on to compare the life-cycle impacts of PVC with other materials used similarly in the built environment, for example aluminum window frames with vinyl window frames, wood siding with vinyl siding, etc. With regard to human health impacts, specifically cradle-through-use, the report states: "aluminum frames are worst among alternatives studied. [With the] addition of end-of-life including burning: aluminum frames remain worst for combined human health impacts, but PVC is worst for cancer-related impacts among alternatives studied." This study included aluminum frames with and without thermal breaks, and considered the energy usage (and methods of energy generation, discussed below) to be as harmful as direct exposure.

Furthermore, the study identified that PVC was worst for cancer-related impacts in piping, siding, and resilient flooring. However, when the life-cycle performance of PVC relative to the other materials was determined, the study questioned whether the focus should be on human health impacts or environmental impacts and where the limits of life-cycle scope are placed. Relative to human health impacts, and a narrow life-cycle cradle-through-use, PVC performed better than some alternatives for windows, siding, and piping, but worst for flooring. However, with end-of-life, and occupational exposures (although the study pointed out that the literature was less complete for occupational exposure data) the study stated "PVC remains among the worst materials studied for human health." In the end, the task groups' recommendations were ambivalent and fractured, relating clear statistics only to stages in the life-cycle and refraining from weighting human health risks with environmental impact since "they found no scientific evidence" to do so.

Primary post-consumer issues with PVC relate to disposal issues. Combustion of PVC produces hydrogen chloride (HCl) due to its chlorine content. The chlorine content is not derived from HCl in the flue gases, but dioxins arise in the condensed solid phase by the reaction of inorganic chlorides with particulate structures in the ash particles. Since most incinerated wastes are washed, dried, and disposed of in landfills, this correlates with the findings relating to increased dioxin levels in monitoring sites adjacent to landfills.

2.3.3 Cement

Typically, cement's impact on global warming is one ton of carbon dioxide for every ton of cement produced. Turning limestone into lime is the main source of carbon emissions. The mined rock must be heated to 2,700 degrees Fahrenheit. Heating the kilns requires enormous amounts of fuel and the most common fuel used in cement plants is coal.

In 2008, about 85 million tons of Portland cement and about 3 million tons of masonry cement were produced at 113 plants in 37 states in the United States. The total cement production capacity was about 130 million tons, most of which was used to make concrete. Approximately 75 percent of cement sales went to ready-mixed concrete producers, 13 percent to concrete product manufacturers, 6 percent to contractors (mainly road paving), 3 percent to building materials dealers, and 3 percent to other users. Texas, California, Florida, Pennsylvania, Michigan, and Alabama were the six leading cement-producing states and accounted for about 48 percent of U.S. production (USGS, 2009b).

Greenhouse gases (GHG) emission reduction strategies by the cement industry include the installation of more fuel-efficient kiln technologies, partial substitution of noncarbonate sources of calcium oxide in the kiln raw materials, and partial substitution of admixtures such as pozzolans, fly-ash, and slag for Portland cement in the finished cement products and in concrete. Because these admixtures do not require the energy-intensive clinker manufacturing (kiln) phase of cement production, their use, or the use of inert additives or extenders, reduces the environmental costs of the cement component of concrete. Fly ash is the by-product of coal-fired power plants. It is estimated that more than 150 million tons of fly ash are produced annually worldwide from the combustion of coal in power plants. However, fly ash typically contains regulated elements such as arsenic, barium, boron, cadmium, chromium, lead, mercury, selenium, etc. With the continued development of fly ash admixture concrete designs, especially in roadway and footings, the potential release of these elements to the environment will generate significant impact to humans, wildlife, and the ecosystem. The environmental impact of coal fly ash has already been studied through leaching tests, and coal fly ash causing groundwater contamination has been reported worldwide. At the present time the Transportation Research Board is studying the environmental impacts of fly ash admixture concrete in roadways.[5]

2.3.4 Glass

The main global impact factor with glass, as with cement, is the production of GHG (primarily CO_2) due to the burning of fossil fuels in the heating of the furnace and production of electricity to supply the compressors. Typically a ton of glass packed will liberate between 500 and 900 kg of CO_2, assuming either a gas-fired furnace or coal-fired electricity usage.

Glass production consists primarily of two types of glass: (1) sheet glass made by the float glass process and used primarily in the built environment, and (2) container glass. Glass containers are wholly recyclable and the industry in many countries has a policy of maintaining a high price on scrap glass (cullet) to ensure recycling rates. The return rates in European countries are close to 95 percent but the return rates are less than 50 percent in the United States and in other countries.

The myth of recycling glass in the building industry is just that. Most recycled glass products are finish materials that are not made from reused windows or mirrors but only contain cullet from container glass. Glass that has been used in the building environment almost always ends up in the landfill. Plate or window glass and mirrors are typically prohibited by recycling plants because these glass products contain contaminants that can seriously damage the furnace or result in poor quality of the plant's product. Glass furnaces operate at temperatures of approximately 2800°F. Lead and aluminum melt at this temperature and the lead settles to the bottom of the furnace tank corroding the brick lining. The aluminum melts into small balls called "stones" or bubbles called "seeds." Stones and seeds can be deposited in the walls of the glass containers being made, causing aesthetic problems and weakening the bottle walls.

2.4 Construction processes and construction equipment used in green building

The construction of green buildings is fraught with inconsistencies between the holistic intent of green building and the actual execution of green building. These inconsistencies frequently stem from incomplete guidelines intended to govern green building construction practices and processes. Green building seeks to mitigate the impact of the building industry on the environment; however, the building industry fails to meet this goal in several ways. The following is a brief discussion of three areas in which there exist inconsistencies between the intent of the governing body and the reality of what is occurring in the building industry today. These areas are:

- Bidding and procurement process
- Construction and demolition waste
- Construction field equipment

2.4.1 Bidding and procurement process

The traditional bidding and procurement process, utilizing the design-bid-build (DBB) project delivery method, occurs only after all construction documents are 100 percent complete. At this point, not only is the design complete but so also are the material and systems specifications. The construction documents are distributed to the general contractor who then distributes specification sections to subcontractors in a piecemeal approach. Little attention is given to the building as a system by either the general contractor or the subcontractors. In essence, the intent of green building design and construction is breached as this point when the holistic systems approach is divided into separate pieces. This lack of attention to the building as a system ultimately leads to inefficiencies and redundancies between the trades, not to mention increased materials waste. Inherent in this traditional approach are time constraints during the bidding process that further minimizes the opportunity for the general contractor to gain full knowledge of the project. Ultimately the general contractor relies on the subcontractors' knowledge of trade-specific systems in a green building project.

Alternative project delivery methods such as design-build (DB), integrated project delivery (IPD), or construction manager at-risk (CM at Risk) involve the contractors during the design process and on through the construction documents phase. During this time contractors can provide valuable insight about construction means and methods; availability of materials specified in local market; and feasibility of proposed recycle programs. More important is the fact that the contractors' involvement during the design process can provide life-cycle cost analysis to facilitate informed decision making by the project's owners. This aligns with the life-cycle cost emphasis in green building and initiates a quality control mechanism that has full project knowledge before construction begins (7group and Reed, 2009; Leffers, 2010).

An alternative project delivery method seems like the obvious choice for a green building project; however, there are obstacles to overcome. The primary problem exists within the guidelines and restrictions by various agencies across the United States that challenge the appropriateness of these alternative methods based on possible acts of collusion or other unethical behavior by the designers and/or contractors. Agency understanding and acceptance of alternative project delivery methods for green building projects is essential to ensure project execution as intended.

2.4.2 Construction and demolition waste

Construction and demolition waste include all forms of waste from the construction, renovation, or removal of built projects such as buildings, roads, bridges, or other non-building structures, and from the clearing of rocks, trees, and dirt. An alternative to traditional demolition is deconstruction. Deconstruction is the 'disassembly' of structures for the purpose of re-using components in the original structure and contributes to reducing the waste stream from construction. The direct reuse of components recovered during deconstruction and used for new construction is the most desirable solution for diverting construction materials from the waste stream. Recycling of materials is the next most desirable solution. Recent estimates indicate that approximately 25 percent of construction waste is recovered for reuse or for recycling purposing. As a result approximately 75 percent, or 107.5 million metric tons, of all construction waste ends up in U.S. landfills each year (Chini, 2007).

Recovery of certain materials is limited by the technologies available for separating and cleaning components of building system. Reuse is often limited by the availability of a recovered material in certain markets. Demand for recovered material is another limitation because without demand there is little incentive to labor through the processes of disassembly, cleaning, and sorting involved with deconstruction. There are cities and states in the United States that have established regulations requiring recycling of a certain percentage of construction and demolition waste. These regulations are not universal and in most areas of the United States are not in place. Until governing agencies establish requirements for the diversion of construction and demolition from the waste stream, the standard practices will continue filling U.S. landfills at an alarming rate of 107.5 million metric tons per year.

2.4.3 Construction field equipment

The construction industry relies on equipment fueled by diesel engines for power generation, excavation, and other site work on a project. This equipment includes generators, backhoes, trenchers, and graders just to name a few. Much of the diesel-powered construction equipment is classified as nonroad vehicles by the Environmental Protection Agency (EPA).[6] Typically these vehicles use red dye diesel. Red dye diesel has the same chemical compound as standard diesel available at any gas station, but has red dye added to provide for a visual distinction between the two types of diesel. The only other difference between the two diesels is that the red dye diesel is less expensive. Red dye diesel is less expensive because it is exempt from state road tax. It is exempt from road tax because it is to be used by nonroad vehicles (Virrueta, 2009).

Diesel powered construction equipment produces nearly forty toxic substances, smog-forming oxides of nitrogen and fine particulate matter, and it contributes to a laundry list of adverse health effects including asthma, cardiovascular and respiratory problems, strokes, heart attacks, lung cancer, and premature deaths in the United States (Scott *et al.*, 2005). The EPA's Clean Air Diesel Rule is a comprehensive national program to reduce emissions from future nonroad diesel engines by integrating engine and fuel controls as a system to gain the greatest emission reductions. To meet these emission standards, engine manufacturers will produce new engines with advanced emission-control technologies.[7] However, the EPA has established a phased approach to accommodate redesign and retooling by manufacturers, along with the replacement by equipment owners. The Clean Air Diesel Rule will be phased in over the next several years. There are estimates that by 2030 requirements in the rule for controlling toxic emissions will be in place and will annually prevent 12,000 premature deaths, 8,900 hospitalizations, and one million work days lost. Until the point at which the restrictions in the rule are required, replacing, refueling, and retrofitting existing equipment is voluntary. Tax incentives exists for owners; however, provisions are also in place to provide relief for equipment owners who demonstrate that adherence to the new rule would cause severe economic hardship.[8]

As previously stated, the intent of green building practices is to mitigate the impact of the construction industry on the environment. Evidence exists that emissions from the diesel-powered equipment necessary to execute a construction project are harmful to both the environment and human population. The Clean Air Diesel Rule includes measures that if implemented by equipment owners will have a significant positive impact and further the purpose of green building. The challenges lie in the extended time for full adoption; conversion of diesel fuel for all nonroad vehicles free of road taxes; and the voluntary status and available relief for owners demonstrating an economic hardship. Eventually the rule will be in force, but in the meantime the health of thousands of people will be directly affected by the negative impact of green building projects using harmful diesel equipment.

2.5 Potential negative side effects of current green building practices

The green building industry is placing heavy emphasis on the reduction of greenhouse gas emissions. Often the GHG reduction focus neglects the regulated industrial pollutants that are also being produced. The U.S. Green Building Council's Materials and Resources points are focused on reducing and off-setting the greenhouse gas emissions created by the manufacture and use of building materials: "Products and materials are climate neutral when there are zero net greenhouse gases, such as CO_2, from the entire life cycle of the product. The manufacturer calculates the total GHG impact utilizing life cycle analysis and then obtains carbon emission reduction credits (ERCs), such as through green power off-set purchases or carbon sequestration projects. The offsets must equal or exceed the GHG produced during extraction, processing, manufacture, transport, and end use of a product, and be certified by a recognized third party using sound scientific and accounting principles" (Scot Horst, Chair, LEED Steering Committee).

At the same time, reducing energy usage by the building sector is recognized as both a security and long-term economic strategy. Unfortunately, the coal-based electrical utility industry is hampering net zero efforts and focusing on maintaining incremental advances in energy reduction goals.

2.5.1 Ground source heat pumps

Ground source heat pumps (GSHP) are universally touted as one of the most efficient electrically powered systems that use the earth's relatively constant temperature to provide heating, cooling, and hot water for homes and commercial buildings. GSHPs can be categorized as having either closed or open loops and those loops can be installed in three ways: horizontally, vertically, or in a pond/lake. For closed loop systems, water or antifreeze solution is circulated through plastic pipes buried beneath the earth's surface. During the winter, the fluid collects heat from the earth and carries it through the system and into the building. During the summer, the system reverses itself to cool the building by pulling heat from the building, carrying it through the system and placing it in the ground.[9]

Depending on burial depth and type of soil, ground loops may be subject to seasonal temperature swings of ±10 degrees Fahrenheit, and in such cases the late-winter, natural ground temperature would drop below 50 degrees Fahrenheit creating conditions that require antifreeze protection. Solutions of alcohols in water – methanol and ethanol – have low viscosity (which translates to lower pumping power requirements) and relatively high heat transfer capability (which translates to shorter ground loops). On the negative side, however, they are highly volatile, flammable in concentrated or pure form, and toxic. Unlike methanol, pure ethanol is not toxic, but only denatured ethanol can be purchased for commercial use as antifreeze. Denaturants render the ethanol toxic and some denaturants will also chemically attack polyethylene piping. Methanol is preferred because it eliminates the possibility of such damage to ground loop piping, and many manufacturers use it as a 20 percent solution.

Additionally, for a given level of antifreeze protection, the cost of propylene glycol is ten times greater than that of methanol, typically in the range of tens of dollars per ton of system heating or cooling capacity, compared with dollars per system ton for methanol solutions.

Methanol is listed by the EPA as a regulated substance, causing health effects even in very low concentrations. Methanol has been banned in certain states as an additive. Studies have shown that methanol spreads and is persistent in shallow aquifers in sufficient quantities (Smith *et al.*, 2003). Other studies are focusing on the potential for groundwater contamination by different organic anti-freeze compounds (ethylene glycol, propylene glycol, and betaine) also used in GSHP (Klotzbucher *et al.*, 2007).

Drilling vertical loops for GSHPs can assume depths of hundreds of feet. Typical housing developments are running loops that are 400 feet deep, with hundreds of houses with units spaced every 100 feet. With groundwater levels typically running less than 200 feet, and the life span of these loops being warranted for less than 25 years, the potential exists of injecting toxic materials into groundwater resources in the name of energy savings.

2.5.2 Carbon sequestration

Combating climate change through the reduction of GHG is a cornerstone to green building. One of the great hopes for combating CO_2 releases is carbon sequestration. The National Groundwater Association is investigating this closely (NGWA, 2008a), monitoring the EPA decisions regarding its regulation and participating in the rule-making process: "The greatest concern with geologic carbon sequestration is the potential risk of leakage of CO_2 or existing saline formation fluids and brines into overlying formations that are underground sources of drinking water, caused by pressure created during injection or by the buoyant nature of CO_2 under typical sequestration conditions. The CO_2 or brine could act to change the chemistry of the underground source of drinking water, causing it to need additional treatment before it can be used. Another concern is that the CO_2 might leak to the surface, causing a potential risk for exposure to humans or the environment. CO_2 is denser than air, so if it leaks into a low area, like a ditch, and cannot mix with the atmosphere it could build up to dangerous concentrations. At very high concentrations (above 80,000 parts per million in the air) CO_2 exposure can cause fatalities." (NGWA, 2008b).

2.5.3 Coal-fired electric plants

Coal is the most common fuel for generating electricity in the United States. In 2007, nearly half (49 percent) of the country's 4.1 trillion kilowatt hours of electricity used coal as its source of energy production. Despite governmental controls, burning coal is a leading cause of smog, acid rain, global warming, and air toxics. In an average year, a typical coal plant generates[10]:

* 3,700,000 tons of carbon dioxide (CO_2)
* 10,000 tons of sulfur dioxide (SO_2)
* 500 tons of small airborne particles
* 10,200 tons of nitrogen oxide (NOx)
* 720 tons of carbon monoxide (CO)
* 220 tons of hydrocarbons, volatile organic compounds (VOC)
* 170 pounds of mercury
* 225 pounds of arsenic
* 114 pounds of lead, 4 pounds of cadmium, other toxic heavy metals, and trace amounts of uranium
* Waste created by a typical 500-megawatt coal plant includes more than 125,000 tons of ash and 193,000 tons of sludge from the smokestack scrubber each year. Nationally, more than 75 percent of this waste is disposed of in unlined, unmonitored onsite landfills and surface impoundments.

- Toxic substances in the waste – including arsenic, mercury, chromium, and cadmium – can contaminate drinking water supplies and damage vital human organs and the nervous system.

In 2007, there were 617 facilities burning coal to generate electricity in the United States. In 2002 the U.S. government sponsored the Clean Coal Power Initiative (CCPI). The CCPI is providing government co-financing for new coal technologies that can help utilities cut sulphur, nitrogen, and mercury pollutants from power plants. Some of the early projects are investigating ways to reduce greenhouse emissions by boosting the efficiency by which coal plants convert coal to electricity or other energy forms. The current CCPI solicitation is focused on developing projects that utilize carbon sequestration technologies and/or beneficial reuse of CO_2 (EIA, 2006).

The Green Building industry is also attempting to address this issue, sponsoring programs such as the Net Zero Energy Commercial Building Initiative, which seeks market transformation of the commercial building sector to net zero energy usage by the year 2030. This initiative is in its infancy and will apply to future buildings, while ignoring the energy usage of existing or other building sectors. Furthermore, there are no plans to sponsor replacement of coal-burning power plants with renewable sources of electrical supply.

2.6 Summary

Evident in the discussion above is the fact that the U.S. green building industry faces challenges in overcoming the inconsistencies between the intent of green building and current standards, codes, and regulations. These inconsistencies affect the very governing mechanisms in place to oversee "green" material production; "green" construction processes; and "green" building occupancy. As previously discussed, one major problem exists in the disposal of byproducts from the production of "green" materials, such as aluminum and cement admixtures, that create harmful byproducts with no real means for safe disposal. Other problems stem from challenges that inhibit the recycling and reuse of construction materials, adding to the growing waste stream. Extended timelines for adoption and conversion from construction equipment emitting deadly toxins is another challenge facing the green building industry. Each inconsistency jeopardizes the wholesale adoption of green building in the United States along with the advantages sought by green building projects attempting to eliminate negative impacts on the environment.

Overall, the lack of a "cradle-through-use" perspective, fueled by a cadre of disconnected piecemeal design and construction practices, contributes to the ongoing inconsistencies. More attention needs to be focused on the ultimate end goal: improved decisions about materials, processes, and buildings – made from a holistic standpoint – that contribute to the elimination of present and future harmful impacts on the environment.

Notes

1 http://www.epa.gov/epahome/aboutepa.htm.
2 http://www.azom.com/Details.asp?ArticleID=3529.
3 http://www1.american.edu/ted/bauxite.htm.
4 http://pvc.usgbc.org.
5 http://rns.trb.org/dproject.asp?n=12636.
6 http://www.epa.gov/nonroad-diesel/.
7 http://www.epa.gov/nonroad-diesel/2004fr.htm.
8 http://www.epa.gov/nonroad-diesel/420f03008.htm.
9 http://www.igshpa.okstate.edu/geothermal/geothermal.htm.
10 http://www.ucsusa.org/clean_energy/coalvswind/c02c.html.

References

7group, and Reed, B. (2009) *The Integrative Design Guide to Green Building*, Hoboken, NJ: John Wiley & Sons.
Altschuler, K., Horst, S., Malin, N., Norris, G., Nishioka, Y. (2007) "Assessment of the Technical Basis for a PVC-Related Materials Credit for LEED," Final Report, USGBC Technical and Scientific Advisory Committee PVC Task Group, February.
Chini, A. (2007) "General issues of construction materials recycling in the USA," in Braganca, L. *et al.* (eds.), *Portugal SB07: Sustainable Construction Materials and Practices – Challenge of the Industry for the New Millennium*, Amsterdam: IOS Press, pp. 848–55.
Energy Information Administration (October 2006) "Coal Production in the United States – An Historical Overview," available online at http://www.eia.doe.gov/cneaf/coal/page/coal_production_review.pdf (last accessed December 26, 2009).
Keeler, M., and Burke, B. (2009) *Integrated Design for Sustainable Building*, Hoboken, NJ: John Wiley & Sons.
Kibert, C. (2005) *Sustainable Construction – Green Building Design and Delivery*, Hoboken, NJ: John Wiley & Sons.
Klotzbucher, T. *et al.* (2007) "Biodegradability and groundwater pollutant potential of organic anti-freeze liquids used in borehole heat exchangers", Center for Applied Geosciences, Institute for Geosciences, Eberhard-Karls-University Tuebingen, Germany, May 2007.
Leffers, R. (2010) *Sustainable Construction and Design*, Upper Saddle River, NJ: Prentice Hall.
National Ground Water Association (2008a) "Comments on U.S. EPA's National Water Program Strategy: Response to Climate Change – Ground Water Protection and Management Critical to the Global Climate Change Discussion" submitted June 2008, available online at http://www.ngwa.org/ASSETS/EC4315A2A315 4E56B003EA6BC50C067D/epa_comments_re_climate_change.pdf (last accessed December 26, 2009).
National Ground Water Association (2008b) "Geologic Carbon Sequestration", September 2008, available online at http://www.ngwa.org/ASSETS/509A6F339 057483E9116F8AFF9B0DB22/info_brief_geologic_carbon_sequestration.pdf (last accessed December 26, 2009).
Scott, J., Silverman, I., and Tatham, S. (2005) "Cleaner Diesel Handbook," available online at http://www.edf.org/documents/3990_DieselHandbook_CleanFuelsRetro fits.pdf , (accessed 11 January 2008).

Smith, L., Molson, J., Maloney, K. (2003) "Potential Impacts on Groundwater of Pure-phase Methanol Releases," Geological Society of America *Abstracts with Programs*, Vol. 35, No. 6, September 2003, p. 525.

U.S. Geological Survey (January 2009a) Mineral Commodity Summaries, Bauxite, available online at http://minerals.usgs.gov/minerals/pubs/commodity/bauxite/mcs-2009-bauxi.pdf (last accessed December 26, 2009).

U.S. Geological Survey (January 2009b) Mineral Commodity Summaries, Cement, available online at http://minerals.usgs.gov/minerals/pubs/commodity/cement/mcs-2009-cemen.pdf (last accessed December 26, 2009).

Virrueta, K. (2009) "Red Dye Diesel – A Red Diesel that Costs Less," available online at http://ezinearticles.com/?Red-Dye-Diesel-A-Red-Diesel-Fuel-That-Costs-Less& id=2721091 (accessed 21 December 2009).

Part II

3 Covenants and building regulations: a twin track approach to improving the energy performance of Dutch buildings

Lorraine Murphy

3.1 Introduction

The Netherlands traditionally enjoyed the term 'front-runner' in terms of environmental policy with early documents such as the 1989 National Environmental Policy Plan recognized as one of the 'first and most comprehensive policy programmes towards sustainable development' (Liefferink, 1998: 86). Alongside this, the Dutch government received frequent praise for promoting sustainable building (Beatley, 2000; Bossink, 2002), for being an early adopter of performance-based regulation and for including a range of stakeholders in policy design and implementation. Creativity and innovation in design and construction have attracted international recognition (Ouroussoff, 2007; Gauzin-Müller and Favet, 2002), as have results of tackling priority areas, such as construction waste (Rovers, 2008). Despite this status and the position of sustainable building on the political agenda for two decades, a common assertion is that it has yet to become mainstream practice (Priemus, 2005; Moss *et al.*, 2005; Van Bueren, 2009). Nearly a decade of debate surrounding how sustainability can be measured means that the environmental chapter of the National Building Decree[1] remains empty. Attracting even more attention is that energy efficiency, the enduring theme of the sustainable building debate, has yet to fully infiltrate the building sector in general and the existing building stock in particular.

Many energy efficiency policies lie outside the statute books, respecting a common Dutch approach to deliberation and self-regulation. A prime example is voluntary agreements, or covenants, as they are known in the Netherlands. Covenants are soft law instruments that can offer strategic support to regulation and form a testing ground for future regulation. As the outcomes of deliberative processes, between government and third parties, covenants embody the characteristic consensus approach to Dutch policy making. Alongside covenants are hard law instruments such as energy standards, which have been enshrined in building regulations in some form or other since the 1960s. With national and international commitments to reduce CO_2 emissions, improve energy efficiency and increase renewable energy, attention is placed on overcoming the long-standing barriers to improving energy performance in the building sector. Whether the Dutch government will concede in favour of stricter, formal regulation or continue to invest in the market and civil society to develop innovative approaches through long-term voluntary processes remains to be seen.

In this chapter the understanding of green, or sustainable building as it is termed in the Netherlands, is presented. The national policy contexts for both sustainable building and the energy aspect of sustainable building are described. The merits and results of building regulations and covenants, two dominant policy instruments directing energy efficiency in buildings in the Netherlands, will be discussed. Opinion of what two decades of sustainable building has brought to the Netherlands is reviewed before conclusions are drawn.

3.2 Sustainable buildings in the Netherlands

3.2.1 Green buildings as sustainable buildings

Green building in the Netherlands is typically conceived of as sustainable building. However, the environmental dimension, and to a lesser extent health aspects, dominate the discussion. Sustainable building is commonly described in terms of minimizing environmental effects of construction, use, renovation and demolition with the positive benefits for health and cost efficiency emphasized (VROM, 2009). Accordingly, the prevailing themes are sustainability of materials, indoor air quality and energy performance. A common target of sustainable building in Dutch policy and academic circles is that the environmental impact of construction be reduced by a factor of 10 to 20 (Rovers, 2008; Van Kasteren *et al.*, 2002). More recently, national and European/international commitments in terms of climate change, energy efficiency and renewable energy have attached exclusive targets to the energy aspect of sustainable building.

3.2.2 A context for sustainable building

The National Environmental Policy Plan (NEPP) of 1989 is frequently credited with launching sustainable building firmly onto the political agenda (Priemus, 2005; Liefferink and Van der Zouwen, 2003; Melchert, 2007). The NEPP is subject to parliamentary approval and sets the legislative framework for Dutch environmental policy. The 1989 version called for full integration of environmental considerations into policy areas (Liefferink and Van der Zouwen, 2003), identified the building sector as an environmental sector (Boonstra and Knapen, 2000) and was one of the first official documents internationally to give serious consideration to climate change (Pettenger, 2007). Documents produced in its wake demonstrated a more consolidated understanding of the environmental impact of the building sector with, for example, the emergence of Life Cycle Analysis (LCA) of materials (Boonstra and Knapen, 2000).

Guidance at national level was offered in the 1990s in the form of National Packages for Sustainable Building. The packages acted as a voluntary checklist for actors in the construction industry. While the packages were widely endorsed by municipalities they also received criticism for the degree to which they truly represented sustainability principles (Van Bueren, 2009). The packages received government support for approximately ten years until a change of government in 2000 led to a shift in policy focus to energy performance and stimulation of demand for sustainable building (ibid). At the time that support was withdrawn, over fifty sustainability rating tools, from both government and private parties, existed on the market (Van den Brand, 2006). The raft of tools drawing on different data, determining sustainability on the basis of different criteria and delivering different outcomes acted as a further miasma to an already complex sustainable building agenda.

In an effort to coordinate the range of sustainability tools, a current project between the government and market parties aims to harmonize a number of tools based on LCA. The outcome of this process will, it is envisaged, be an environmental performance standard for buildings that will be incorporated into the Building Decree during its next proposed revision in 2011. Such a move would finally furnish the environmental chapter of the Building Decree with some content.

The 1989 NEPP and supplementary documents not only established the content and scope of policy in terms of sustainable building but also the style (Liefferink and Van der Zouwen, 2003). Coinciding with a governance shift across many western societies the NEPP reflected a growing scepticism towards direct regulation and prescriptive technical solutions (Bressers and de Bruijn, 2005). An approach designed to encourage strategic relationships with stakeholders and self-regulation was increasingly viewed as more effective than a distant and authoritarian policy style (ibid). An expression of this governance shift is the covenant process, a defining characteristic of how the Dutch government interacts and shares responsibility with stakeholders in the development and implementation of environmental objectives. Traditional regulatory approaches remain intact, however, with building regulations at the helm in establishing minimum standards.

Alongside building regulations and covenants the Dutch government uses a range of carrots, sticks and sermons to stimulate change (Bemelmans-Videc, *et al.*, 1998). Economic instruments play a role as carrots overcoming market failures when it comes to the promotion of sustainable products, while subsidies and green mortgages serve to increase the financial attractiveness of sustainable building. Sticks in the form of a ban on construction waste going to landfill have brought sustainability principles to different parts of a building's life cycle. Demonstration projects have acted as sermons to dispel critical voices by showing that sustainable buildings do not necessarily diverge from conventional Dutch building design, need not entail excessive costs and can stimulate consumer demand. Information tools stemming from NGOs, government and energy suppliers have increased in sophistication with a range of interactive web-based tools available. Tailored energy advice for homeowners form more personalized information campaigns. More recently, instruments have adopted an international flavour with the Dutch Green Building Council launching BREEAM NL for new buildings in 2009 (DGBC, 2009).

While instruments developed over the last number of decades reflect the spectrum of sustainability issues, the energy aspect has invariably dominated. Energy remains the only sustainability aspect regulated in the Building Decree. As currently formulated it is the theoretical energy use in the user phase of a building that receives regulatory attention. As a result energy embodied in construction products or prevalent at different stages of a buildings life cycle remains untouched by building regulations. Neither is energy positioned within the wider sustainability context representing a missed opportunity in achieving concomitant gains for aspects such as water efficiency.

3.2.3 A context for energy efficiency in buildings

Since the energy crisis in 1973, energy has outperformed other sustainability issues in terms of information campaigns, economic incentives and regulations. The context for energy efficiency has altered over the years with the current government framing action in terms of energy security, climate change and economic competitiveness (VROM, 2007). These three issues are tackled in one process known as the energy transition. As with sustainable building over a decade earlier the energy transition received its national political debut through the NEPP. In 2001, in the fourth NEPP, it was argued that intensifying policy instruments for environmental problems like climate change would be inadequate; instead, "solving the major environmental problems requires system innovation; long drawn-out transformation process comprising technological, economic, social-cultural and institutional changes" (VROM, 2001: 30).

In 2007 the Dutch cabinet launched the 'Clean and Efficient' (*Schoon en Zuinig*) work programme to strengthen the energy transition. National targets encompassing, and at times going beyond, EU climate and energy package targets and Kyoto targets were proposed, such as:

- a reduction of greenhouse gas emissions by 30 percent (baseline year 1990)
- a reduction of energy consumption by 2 percent per year by improving energy efficiency
- an increase in the share of renewable energy in Dutch energy consumption to 20 percent, with all three targets to be achieved by 2020 (VROM, 2007: 3)

With responsibility for approximately 37 percent of total energy use in the Netherlands (cited in Itard *et al.*, 2009: 3) it is not surprising that the building sector has become a focus for realizing these ambitious targets. The energy-saving potential associated with buildings is demonstrated in the Dutch response to the European Energy Services Directive (ESD).[2] The ESD stipulates that member states achieve a 9 percent improvement in energy efficiency by 2016. The building sector features as the largest contributing sector to the Dutch ESD target (Ministry for Economic Affairs, 2007). The central role afforded to this sector is not unique to the Dutch perspective. The EU recognizes the building sector as the largest energy user with responsibility for the majority of CO_2 emissions. Legislation such as the Energy Performance of Buildings Directive (EPBD)[3] is used to lever cost-effective savings which are viewed to hold the potential of reducing final energy use in the EU by 11 percent (EC, 2008: 2).

Box 3.1 details measures used to stimulate energy efficiency in the Dutch building sector. For new buildings, regulatory attention focuses on tighting current performance standards as established by building regulations. Existing buildings have limited exposure to performance standards and the Energy Performance Certificate (known as the Energy Label in the Netherlands) as required under the EPBD is instead the main regulatory focus. Three national 'energy performance' covenants play a role, assigning specific targets to existing buildings, the social housing sector and new buildings.

BOX 3.1 POLICY TOOLS SUPPORTING ENERGY PERFORMANCE IN BUILDINGS IN THE NETHERLANDS

Existing buildings

- **Energy Label:** designates an energy rating (A–G) for buildings. Investigations into the legal implications of requiring all buildings on the market to rate C or higher is planned.
- **Building Decree:** sets thermal envelope requirements for buildings undergoing extension/renovation. Standards for new build are required for complete renewal.
- **Covenant 'More with Less':** aims to overcome the barriers to improving energy performance in existing houses. Targets include improving the performance of 500,000 houses by 2011 and 300,000 per year until 2020.
- **Covenant 'Energy Saving in the Corporation Sector':** aims to stimulate energy saving in the social rental sector (35 per cent of dwellings in the Netherlands).

- **Social Rent Rating System**: is due to include energy as a factor dictating rental charges for social housing. Energy as a quality aspect will be inferred from the Energy Label.

New buildings

- **Energy Performance Co-efficient (EPC)**: will be revised with aims of achieving:
  - energy neutrality in residential buildings by 2020
  - 50 per cent improvement in energy efficiency of utility buildings by 2017
  - Government buildings achieve energy neutrality by 2012
- **Covenant 'Spring Agreement-Energy Saving in New Build'**: aims to achieve energy neutral buildings by 2020. Objectives include the development of a revised energy performance methodology to achieve this target.

Crosscutting policy tools

- **Subsidies**: promote advice and innovative technologies. Schemes from national and local governments, banks and energy suppliers operate often over short time periods.
- **Information tools**: provide information on subsidies, policy support and compliance checks for legislation. Milieu Centraal, COEN, HIER and branch organisations are the main active organisations.
- **Energy Performance of Buildings Directive**: mandates the Energy Label, stipulates minimum standards for new buildings and existing buildings under renovation and requires inspections on installations such as boilers and air conditioning units.
- **European Eco-design Directive**: regulates efficiency improvements in energy using products and acts as a key component of the legislation package for buildings.

Adapted from Ministry of Economic Affairs, 2007; VROM, 2007.

While building regulations and covenants represent enduring features of the policy landscape, a range of additional tools act as further bait for energy performance gains. It is proposed that the national Social Rent Rating System will include energy as a factor determining rental charges. This is an attempt to overcome the barrier of 'split incentive', meaning that social housing landlords could theoretically recoup the costs of investing in energy efficiency through demanding higher rents. Information and economic tools are introduced at different stages of energy efficiency campaigns to increase awareness and reduce market inequalities in terms of sustainable energy products and services. In addition, implementation of the European Eco-Design Directive[4] focuses on the energy used in buildings in terms of product use, complementing the EPBD's focus on the building envelope.

Nonetheless, two tools varying significantly in content, scope, actor involvement and legal basis typify the energy aspect of sustainable building in the Netherlands: one hierarchical and traditional based on public law expressed chiefly through energy standards in building regulations, and the other based on private law and voluntarism between government and market parties through covenants (Glasbergen, 1999; Van der Waals and Glasbergen, 2002). Often building regulations and covenants are explicitly linked (Glasbergen, 1999; Zito *et al.*, 2003) as covenants bolster regulations identifying avenues for more innovative approaches. How building regulations and covenants are organized and enforced and results of implementation are described in the following sections.

3.3 The Building Decree and energy performance regulations

3.3.1 Evolution and organization of the Building Decree

The backbone of energy efficiency legislation in the Netherlands is the National Building Decree and associated energy performance regulations. The decree was developed from national standards first introduced in 1965 by the Association of Dutch Municipalities (Van der Heijden *et al.*, 2007). The result was the Model Building Bye-Law,[5] which was not mandatory but was widely adopted. In terms of energy, the model initially prescribed thermal insulation standards but over the years its influence extended to prescribing requirements for components such as double-glazing (Van Cruchten *et al.*, 2008).

A desire for a centralized, uniform and performance-based suite of regulations resulted in the introduction of the National Building Decree in 1992 (Visscher and Meijer, 2008; Ang *et al.*, 2005) with a current version dating from 2003. The legal basis for the decree is the Housing Act.[6] Technical detail is contained in the associated regulations while standards developed by the Dutch Standardisation Institute can verify that established values are attained (Visscher and Meijer, 2006).

The Building Decree consists of five chapters covering health, safety, usability, energy use and environment. The content and scope of the environment chapter is yet to be developed, however. Proposals to revise the decree in 2011 and a current project focused on harmonizing a number of LCA tools should witness the inclusion of performance criteria for materials in the environment chapter.

Energy aspects are dominated by the Energy Performance Norm (EPN), which represents an integrated method for calculating energy use in a building. The result of the EPN is the Energy Performance Co-efficient (EPC) defined as 'the characteristic energy use of a building divided by the standardized energy use' (cited in Beerepoot, 2002). Characteristic energy use is understood to be space heating, hot water heating, lighting and energy use by fans, cooling and humidification installations (ibid). As well as the performance requirement for the whole building, the decree contains several basic requirements including access to daylight and minimum standards for insulation, air permeability and ventilation (Rovers, 2008).

A key criterion in the design of the performance-based regulations was that design freedom and innovation would not be constrained, therefore technical solutions are avoided (Visscher and Meijer, 2008; Ang *et al.*, 2005). This allows freedom in deciding how elements can be combined to reach the EPC value. There are no qualification or training requirements for calculating the EPC and it is not linked to an accreditation process. Calculations demonstrating that the EPC has been achieved are submitted to municipality building control departments as part of the building permit process.

The EPC is the stalwart component of the Building Decree. It has been incrementally tightened over the years in a quest to continually improve overall energy efficiency and to stimulate innovation. In 1995 the EPC for a dwelling was 1.4, which at the time represented an improvement of about 10 percent in energy use (Rovers, 2008: 24). Currently, the EPC is 0.8 with proposals to reduce this to 0.6 in 2011 and to 0.4 in 2015 with the aim of reducing fossil fuel use by 50 percent (VROM, 2007: 24).

3.3.2 Compliance and enforcement

While the formulation of building regulations is centralized, compliance and enforcement arrangements remain with municipalities. An amendment to the Housing Act in 1992 permits the transfer of regulatory supervision to regional authorities or the private sector (Van der Heijden *et al.*, 2007) although municipalities largely maintain their traditional duties (Van der Heijden, 2009). Compliance and enforcement is organized around the planning and construction phases of a building project with no provision for post construction monitoring and verification (Visscher and Meijer, 2008). This exposes particular problems for energy performance as it is strongly dictated by building practice (Joosen, 2007).

It is claimed that the compliance rate with building regulations is approximately 70 percent in the Netherlands, a figure that is in line with the average in other IEA countries (IEA, 2009: 41). Nonetheless, other sources indicate that the actual rate is somewhat lower (see Van der Heijden, 2009). Furthermore, research focusing on compliance with the energy aspects of building regulations presents a starker picture. A study commissioned by the Ministry of Environment found that in 154 new house developments 25 percent contained incorrect EPC calculations while the EPC upon final construction was incorrect in over half the cases (Kuindersma and Ruiter, 2007).

Lack of compliance and enforcement of building regulations in general and energy standards in particular is explained by a range of factors including:

- a lack of responsibility on behalf of actors in the building industry
- municipalities not fulfilling their control duties
- low priority attached to energy aspects of regulations
- complexities attached to verifying the EPC calculation
- differences between the EPC value issued in the building permit and that of the final constructed product
- the rare use of sanctions (Visscher and Meijer, 2008; Energids, 2006; Joosen, 2007).

3.3.3 Impact of energy performance regulations

Despite difficulties with respect to compliance and enforcement, energy standards are viewed to have yielded gains. The most oft-cited positive impact of building regulations in terms of energy use in the Netherlands relates to product development and dissemination. That gas condensing boilers are the norm and insulation products and high-performance glazing have improved are attributed to the tightening of EPC values (Beerepoot, 2007; Joosen, 2007). Accordingly, it is stated that contemporary houses use approximately half the energy for heating as houses constructed in the 1970s (cited in Van Bueren, 2009). Nonetheless, Beerepoot (2007) argued that tightening of EPC values led to incremental advancements such as product improvements and to the diffusion of existing technology. More innovative measures such as solar thermal systems or heat pumps have not been supported through this policy stance (ibid).

While the effects of building regulations are not systematically monitored and evaluated, the Dutch government and interested parties have commissioned several research studies to judge performance. The most recent, a statistics-based study commissioned by parties to the Spring Agreement examined the effects of tightening EPCs on energy (gas) use (Itard *et al.*, 2009). Results showed no correlation between energy use and more recent tightening of EPC values finding that houses constructed in the period of 2000–2006 have the same energy use levels as those constructed in 1998–1999. While the EPC had an initial affect on energy use it does not appear to have been sustained during subsequent tightening of the values. The results of this study suggest that the EPC has reached an optimum level with the role of the user and actual energy use now having attained greater importance than that of theoretical energy use. The policy direction of reducing the EPC value further is questioned by the Itard *et al.* (2009) study with suggestions that focus should be placed on:

- means of securing the actual realization of calculated performances
- procedures for control and monitoring of HVAC equipment
- research on actual energy use in relation to behaviour especially in relation to increased electricity use
- possibilities for integrating regulation with robust concepts such as that associated with Passive House.

While the preceding sections dealt with some issues relating to the content and compliance of regulations an additional issue is one of scope. Regulations typically apply to new constructions. The reality that a low-carbon building stock in this century depends on a stock largely constructed before the advent of the most modest of energy standards moves the spotlight onto existing buildings.

3.3.4 Existing buildings

Existing buildings are largely immune to energy performance regulations in the Netherlands. Extension developments and renovations requiring a building permit must meet basic standards for the thermal envelope (Itard and Meijer, 2008). The principle of acquired historical rights means that ambitions for existing buildings are protected by the regulations in place at the time of original construction (Ang *et al.*, 2005). The result is that innovative means of tackling existing buildings, such as requiring consequential works to upgrade the energy performance of a whole building during renovation/extension, are not used in the Netherlands. Typically soft measures have been used to target the existing stock, one of the most structured being the voluntary scheme known as Energy Performance Advice (EPA). EPA consisted of an energy scan with tailored advice by a trained inspector representing a precursor to the Energy Label system of today.

In theory, the most far-reaching regulatory influence on existing buildings in the Netherlands is the Energy Label. According to the EPBD an Energy Label is required at trigger points such as the sale and rental of a building making the existing building stock accessible. The idea behind a mandatory Energy Label is that with energy rating communicated, higher-rated buildings will have a market advantage. The legal ramifications of requiring all buildings on the market to reflect a minimum C rating is being investigated by the Ministry for the Environment (VROM, 2007). Despite the significance afforded to the Energy Label in policy documents, particularly to the role of this instrument in reaching the more elusive existing housing stock, there are significant issues surrounding this instrument in the Netherlands.

3.3.5 The Energy Performance of Buildings Directive and the Energy Label

The EPBD is regarded as a cornerstone policy in the EU's energy efficiency and climate change campaign (ENDS Europe, 2005). Its introduction in 2002 established a number of requirements for member states; specifically,

- a methodology to calculate integrated energy performance of buildings
- minimum energy performance standards for new buildings and major renovations
- certification of buildings when constructed, sold and rented and for all public buildings
- inspections of boilers and air-conditioning systems
- certified/qualified experts for the certification and inspection processes.

With performance regulations embedded in the building regulations since 1995, and voluntary certifications for existing housing under the EPA scheme, transposition of the EPBD did not immediately appear to represent a major challenge for the Netherlands. The mandatory certification of buildings, however, became the most controversial aspect of the directive and the following discussion focuses on this aspect alone.

Article 7 regarding mandatory certification procedures was one of the few aspects of the EPBD that required legislative change in the Netherlands. As a result, the Energy Performance of Buildings Decree[7] and Regulation on Energy Performance of Buildings came into force in December 2006 (Van Ekerschot and Heinemans, 2008). The original transposition date of 4 January 2006 for the EPBD passed as the Netherlands exploited the three-year grace period allowed to ensure the availability of experts for the certification process. Eventually, Energy Labels became mandatory from January 2008. However, the slow transposition rate was not due only to a lack of certified experts. The Dutch government announced in 2005 that the directive would not be implemented due to the administrative burden (ENDS Europe, 2005). The move was viewed as unprecedented at the time and a challenge to EU legal procedures (ibid). While the position was eventually recanted, the objective to minimize costs and the administrative burden defined the Energy Label system from the outset.

The introduction of Energy Labels spurred immediate criticism in terms of: quality of content, transparency, calculation methodology and adequacy of supporting administrative structures. In terms of content, recommendations on how to improve the energy efficiency of buildings were not a standard component of the Dutch Energy Label, despite this being urged by the European Commission. Initial discrepancies between Energy Labels of apartments in the same building damaged credibility. Additionally, companies were accredited to carry out the Energy Label, which did not guarantee that inspectors received the relevant training. In terms of transparency, customers were not permitted access to the data dictating the energy rating. Issues surrounding the qualification of inspectors, transparency, and the lack of an adequate complaints procedure led the Dutch Home Owners Association to advise homeowners to ignore the regulation (Vereniging Eigen Huis, 2007).

A further issue was that labels were introduced without a dedicated compliance and enforcement procedure but were to be governed by the market. While in some member states parties involved in the transaction process, such as notaries, stepped up as quasi enforcers, the same has not generally occurred in the Netherlands. The absence of a label does not entail a sanction and common practice is that parties involved in the transaction agree that a label is considered unnecessary. If such agreement is not reached, the affected party can appeal under the Dutch Civil Code.[8]

Added to this the Energy Label sits precariously within the wider building regulatory system. The EPBD requires that labels be available upon construction which is circumvented in the Dutch case with the interpretation that the EPC fulfils this requirement. This is despite the fact that the EPC process is not associated with an accredited procedure and does not form a communication instrument.

As a result of the myriad of difficulties, the Energy Label experienced a humble beginning with one study finding that in the first nine months of 2008 less than 20 percent of a sample of transactions was associated with a label (Brounen *et al.*, 2009). This study showed geographical basis with labelled dwellings typically located in areas where the housing market is less competitive. (Though demonstrating problems with the take-off of Energy Labels this study did find that houses with a higher energy rating held a marginal market advantage supporting the fundamental reasoning behind the tool).

A revised Energy Label was introduced in January 2010 in response to a number of criticisms in terms of quality and methodological aspects. However, a formal compliance and enforcement component remains to be developed. Interestingly, despite the unforgiving treatment of this instrument in its early stages, this tool is gaining increasing attention in terms of its malleability for wider policy objectives. In stakeholder discussions over the merits of following an obligation-driven rather than an incentive-driven approach to the housing sector, the Energy Label is emerging strongly in terms of the potential of making certain Energy Label ratings mandatory.

The route of the EPBD in the Netherlands reflects a common understanding of the relationship between European and Dutch policy. Liefferink and Van der Zouwen (2003) note that policy instruments emanating from the EU rarely represent a misfit with Dutch policy tools and typically instruments are already in place in the Netherlands in some form. This is confirmed by experience with the EPBD which was introduced at a time when performance-based regulations and labelling systems (though voluntary) were already established. Liefferink and Van der Zouwen (2003) add, however, that the EU's legalistic and formal policy style is sometimes at variance with the Dutch approach of negotiation, consensus and long-term planning. The Dutch method of investing in the energy transition as a means of overcoming barriers to achieving energy performance improvements and of tackling the existing housing stock with soft measures such as voluntary labels are examples of this juxtaposition. The three national energy performance covenants (see Box 3.1), focusing on existing houses, new build and the social sector, are even more apt examples of this policy style.

3.4 Covenants and the energy transition

3.4.1 Evolution and organization

Since the 1980s, covenants have come to occupy a central tier of Dutch policy representing 'a major shift in the philosophy of environmental governance and regulation' (Bressers and De Bruijn, 2005: 276). Disenchantment with direct regulation led to support for processes of negotiation espousing private initiative and shared responsibility (ibid). Covenants can be understood as agreements over a set time period between government and sector representatives, typically branch organizations. In this way covenants can form a platform where national targets are translated into digestible sector-specific portions negotiated by relevant parties. In the early days covenants were more akin to gentleman's agreements between industries and government and focused on single issues (Glasbergen, 1998). Contemporary covenants expand the range of target groups and include quantitative goals, designated responsibilities, mechanisms for monitoring and evaluation as well as agreed costs and terms. The European Commission adds another requirement to ensure the effectiveness of agreements; that is, sanctions in the event that targets are not met (cited in Bertoldi and Rezessy, 2007).

It is stated that a covenant has no formal place in environmental law but instead functions as a supplement to and articulation of legal frameworks (Glasbergen, 1998; Bertoldi and Rezessy, 2007). Winsemius (1993: 6) places covenants 'somewhere between indirect regulation and self regulation'. The Dutch government provides some further grounding stating that covenants should not take precedence over formal legislation but be restricted for use:

• in anticipation of regulation – as an interim measure
• if regulation is likely to become redundant
• if different forms of regulation need to be explored
• to support regulation (Bressers and De Bruijn, 2005).

The research of Bressers and De Bruijn (2005) found that covenants are also used to highlight an issue, therefore acting as a symbolic policy instrument. The above criteria find sympathy with challenges of seeking improved energy performance in the building sector. The energy transition recognizes that simply strengthening or adding new regulations may be inadequate to reverse the unsustainable patterns of energy use that are embedded in the structures and institutions of Dutch society (Loorbach *et al.*, 2008). The three national 'energy performance' covenants (see Box 3.1) reflect the aims of exploring alternative instruments and supporting regulations. Additionally, the image presented of businesses, research organizations, government and NGOs working together in search of agreeable and influential results finds some symbolic sentiment.

3.4.2 Compliance and enforcement

Given that covenants are fundamentally soft measures, associated compliance and enforcement regimes can be fragile. Covenants operate through rules of conduct and business codes rather than through compliance procedures and sanctions (Glasbergen, 1998). On the one hand, covenants theoretically succeed where direct regulations fail in harnessing the problem-solving abilities of relevant parties towards goals more ambitious than minimum standards. On the other hand, covenants can be marred by a lack of commitment and inaction with limited recourse for redress. Notable characteristics of covenants are first, parties are not necessarily legally bound to targets and second, commitment is typically received from one level of a stakeholder group. In terms of the first point, out of the three national covenants formulated to realize energy efficiency objectives in the building sector only 'More with Less' includes a statement that the agreement is binding under civil law. All three covenants state that parties are bound to the covenant for its duration but that this is not legally enforceable. In terms of the second point, parties to covenants are typically branch organizations who in reality have no hierarchical authority over the parties they represent (ibid).

The lack of legality associated with covenants gives rise to divergent opinion. Some commentators view this as a key strength with parties less objects of regulation and more active participants in a process geared to environmental change (ibid). Providing covenants with more precision in legal terms would, in the opinion of these commentators, annihilate the key advantage of flexibility (ibid). Even loyalty to the term of covenant instead of voluntary agreement in the Dutch context is said to pertain to the hints of legal obligation associated with the term 'agreement' (Zito *et al.*, 2003).

On the other hand Bertoldi and Rezessy (2007) believe that essential elements of a covenant are credible and enforceable sanctions. An alternative to such sanctions is the presence of a threat, such as regulation in the event of inaction (ibid). This point of view is supported by Biekart (1998), who views the threat of regulation and legally binding targets as necessary for the success of covenants. These characteristics are not immediately evident in the three national covenants guiding energy performance improvement in the Dutch building sector.

3.4.3 Impact of covenants

Like many second-generation policy instruments, covenants suffer a dearth of ex-post analysis. There is a further marked absence of analysis on how covenants interact with other policy instruments even though this interaction is viewed as a key raison d'être of the instrument. The consistent absence of demonstrable results means that a raft of covenants are considered entirely ineffectual and as a delay to real action (Golub, 1998).

A qualitative evaluation of the three covenants developed to stimulate energy performance improvements in the building sector in the Netherlands was conducted in 2010 (Schneider and Jharap, 2010). This evaluation highlighted some important developments and proposals attributed to the covenants, e.g. the proposal that social housing rental mechanisms include energy performance (as reflected in the Energy Label) as an aspect dictating rent in the future (ibid). Furthermore, the covenant for the social housing sector has resulted in action by housing associations, many of which are now developing and aligning energy policies to covenant targets. In terms of quantitative information on whether covenant aims are being met, there is, however, less evidence. Furthermore, while it is early days for the covenants, many of the criticisms from literature are evident in the practice of these covenants in the Netherlands. Chief among these criticisms is whether the voluntary commitment of stakeholders will be adequate and consistent enough to meet the 2020 targets for which they share responsibility.

Some commentators argue that the impact of covenants should not be searched for in quantitative terms but in process terms. Bressers and de Bruijn (2005) note that covenants raise the level of ambition of policy, changing attitudes of target groups and improving channels for collaboration between the government and sectors. Covenants are also found to improve knowledge among stakeholders (ibid). The potential benefits of the three national covenants in developing much-needed knowledge and collaboration in a unique gathering of groups like home-owners associations, renter associations and estate agent representatives may yet be a future positive impact.

Returning to the Dutch government's criteria for choosing covenants over formal regulation is an understanding that covenants are best used in a certain phase of a policy cycle. This phase is characterized by exploration when potential solutions are researched, tested and developed. Bressers and de Bruijn (2005) interpret covenants as a step towards stronger regulation. They believe that when solutions are presented the question is justified whether the covenant should be succeeded by regulation. It remains to be seen whether the three national covenants for energy performance improvement result in a revamped suite of regulations.

3.5 Sustainable building in the Netherlands

Despite occupying a niche for twenty years, it is widely accepted that sustainable building is not yet mainstream practice in the Netherlands (Rovers, 2008; Van Bueren, 2009; Priemus, 2005; Moss *et al.*, 2005). Early opinion emphasized that the environmental impact of building should be reduced by a factor of 10 to 20 (Rovers, 2008). Yet is it stated that leading building projects have an environmental performance of approximately a factor of 2 better than ten to fifteen years ago (ibid). Energy as the mainstay of the sustainable building debate has shown some mixed successes with the building sector remaining as the largest unused profitable potential for energy saving (cited in Van Bueren, 2009). This gap between high ambitions and actual implementation is interpreted in a number of ways by proponents, some of which are described in more detail below.

Priemus (2005) argues that a major challenge to success lies in how sustainable building is conceptualized. The definition of sustainable building is considered overly broad and the absence of an agreed measurement system defies evaluation of progress. The prolific growth in sustainability rating tools using different criteria and defining different outcomes is testament to the complexity of the issue. The seemingly intractable challenge attached to operationalizing a version of sustainable building on which some level of consensus has been achieved is evident in the environmental pillar of the Building Decree existing in title form only for nearly a decade.

While conceptual and operational issues may have delayed progress, Moss *et al.* (2005) view a major issue to be the perception of what the barriers to sustainable building really are. Often a linear interpretation of challenges focuses on lack of funding, inadequate information and weak regulation. The result is a resource focus on these three issues when in reality a range of policy agendas and actor interests need attention (ibid). This view is supported by Van Bueren (2009), who views the bottleneck in promoting sustainable building as caused by the institutionally fragmented context of design, construction, management and maintenance in the built environment.

Besides issues with the ambiguity of concepts and sectoral issues, there remains a central challenge of developing a suite of instruments that can comprehensively tackle sustainability in the building sector. A focus on energy, as the most developed aspect of the sustainable building agenda, shows the difficulty attached to developing a complementary suite of policy instruments that capture the strengths and overcome the weaknesses of individual tools. Energy performance regulations demonstrated early successes resulting in reductions in the amount of energy for heating buildings and product development. Nonetheless, current regulations display weaknesses when it comes to capturing higher ambitions. Moreover, despite the security afforded this public law instrument, research identifies significant issues with compliance and enforcement across the range of actors involved.

Covenants offer the alternative of extending beyond minimum standards without the time-consuming and legalistic shackles associated with regulations. However, while boasting informality and flexibility as key strengths, there is as yet little evidence that covenants will lead to swifter or improved results in achieving energy performance improvements in the building sector.

Both regulations and covenants struggle to bring existing buildings, and in particular the existing private housing stock, into the fold. In this critical subsector the sanctity afforded to the private home-owner dictates that soft law instruments remain the preferred approach. The EPBD offered a public law entry point with the mandatory Energy Label but this tool received an initial luke warm response by the Dutch government and interested parties. Instead, labels came into conflict with the deregulation agenda and were diluted in a process of minimizing and optimizing regulation. As a result, the perceived benefits that spurred their introduction at the European level so far have been overshadowed. Whether the revised Energy Label introduced in 2010 and the proposed integration of this instrument with other mechanisms in the property market can reverse this trend remains to be seen.

It is widely accepted by commentators that, to move the sustainable building agenda forward and to realize ambitious CO_2 emission reduction targets, existing measures need to be strengthened and expanded and new approaches tested and developed across the sectors (Opstelten *et al.*, 2007; Energy Transition Task Force, 2006; Rovers, 2008; VROM, 2007). With the energy transition there is acceptance that a low-carbon society will involve "a long drawn out transformation process" (VROM, 2001: 30). The battle of developing a comprehensive package of instruments that are consistent and reinforcing and that engender the right balance between carrots, sticks and sermons looks set to continue.

3.6 Conclusions

Over the last two decades a policy context, to varying degrees of strength, has been forged for sustainable building in the Netherlands. A number of achievements such as the early adoption of performance-based regulation and inclusion of third parties in the policy process earned the Netherlands a front-runner status. Over the years the complexity of the sustainable building agenda has stymied action perhaps most symbolically evident by the empty environmental chapter of the Building Decree. Moreover, pressing targets, especially in terms of energy performance, has exposed weaknesses in the approach and instruments used. There is acceptance, iterated by government departments, research organizations and NGOs, that the ambitious targets to combat climate change and energy inefficiency issued in national documents and in response to EU policy need a corresponding suite of ambitious tools. Alongside this, barriers to accessing the existing building stock persist and the reserve of environmental gains, in particular in terms of energy use, remain untapped.

Two tools in particular operate in a twin-track approach towards achieving goals for energy efficiency, renewable energy and CO_2 emission reduction in the building sector. Covenants and building regulations, occupying polar ends of a legal spectrum, are central efforts for delivering targets. A scan of how these tools work in practice uncovers a number of weaknesses. Ad hoc monitoring of regulations show that outcomes do not necessary measure up to the ambitious aims assigned to them. Differences between calculated and actual energy performance and uncertainties related to behaviour and maintenance of installations undermine confidence that actual energy use is currently being tackled by regulations. Debates over the costs and function of Energy Labels have overshadowed the promise of this tool in reaching the more evasive elements of the sector, namely the existing privately owned stock. Problems with enforcement of energy performance regulations and the absence of a regime for Energy Labels further shake confidence. Meanwhile covenants represent a long-term and strategic view towards achieving targets in a more informal multi-actor setting. However, covenants depend greatly on the moral commitment of stakeholders, the absence of which places the complete process at risk of failure.

The energy transition provides space to scrutinize the instruments used to develop more sustainable energy systems across the sectors. It remains to be seen if this transition approach can be a hotbed of innovation proposing instruments suitable for the targets they seek to achieve. At the same time current climate change science demands that the energy transition must be swift with predictions that a delay in abatement action by ten years will make it virtually impossible to keep global warming under 2°C (McKinsey and Company, 2009). Proponents argue that government support through a stronger legislative regime is the swift response that is required. Enthusiasts for covenants emphasize the role of this instrument in supporting and further developing this legislation. In theory the result is a promising combined approach. In practice much depends on multi-actor participation that is voluntary, and on the design and enforcement of a public law instrument that is responsive to demanding targets. In the Netherlands, as elsewhere, a central challenge is putting tools on the right track and ensuring that these tools can be strong individually and collectively.

3.7 Postscript

In October 2010 a new government was installed in the Netherlands. The new government has established that previous climate change targets for 2020 will be reduced. Namely, the greenhouse gas emission target of 30 percent will be reduced to 20 percent and the share of renewable energy target of 20 percent will be reduced to 14 percent.

Acknowledgements

Thanks to Prof. Henk Visscher, Dr. Frits Meijer, Dr. Neil Murray and Dr. Jeroen van der Heijden for comments on drafts of this chapter. Thanks to Dr. Arjen Meijer for information on the proposed environmental chapter of the Building Decree.

Notes

1 National Building Decree, Bouwbesluit 2003.
2 Directive 2006/32/EC of the European Parliament and of the Council of 5 April 2006 on energy end-use efficiency and energy services and repealing Council Directive 93/76/EEC [2006] OJ L114/64.
3 Directive 2002/91/EC of the European Parliament and of the Council of 16 December 2002 on the energy performance of buildings [2003] OJ L1/65.
4 Directive 2005/32/EC of the European Parliament and of the Council of 6 July 2005 establishing a framework for the setting of ecodesign requirements for energy-using products and amending Council Directive 92/42/EEC and Directives 96/57/EC and 2000/55/EC of the European Parliament and of the Council.
5 Model Building Bye Law, Model-bouwverordening (MBV) 1965.
6 Housing Act, Woningwet 2003.
7 Energy Performance of Buildings Decree, Besluit energieprestatie gebouwen 2006.
8 The Dutch Civil Code, Burgerlijk Wetboek.

References

Ang, G., *et al.* (2005) "Dutch performance-based approach to building regulations and public procurement", *Building Research and Information* Vol. 33 No. 2, pp. 107–119.
Beatley, T. (2000) *Green Urbanism: Learning from European Cities*, Washington: Island Press.
Beerepoot, M. (2002) *Energy Regulations for New Buildings: In Search of Harmonisation in the European Union*, Delft: DUP Science.
Beerepoot, M. (2007) *Energy Policy Instruments and Technical Change in the Residential Sector*, Amsterdam: IOS Press.
Bemelmans-Videc, M. L., *et al.* (1998) *Carrots, Sticks and Sermons: Policy Instruments and their Evaluation*, New Brunswick, NJ: Transaction Publishers.
Bertoldi, P. and Rezessy, S. (2007) "Voluntary agreements for energy efficiency: review and results of European experience", *Energy and Environment*, Vol. 18 No. 1, pp. 37–73.
Biekart, J. W. (1998) "Negotiated agreements in EU environmental policy, in Golub, J. (Ed.), *New Instruments for Environmental Policy in the EU*, London: Routledge, pp. 165–189.
Boonstra, C. and Knapen, M. (2000) "Knowledge infrastructure for sustainable building in the Netherlands", *Construction Management and Economics*, Vol. 18 No. 8, pp. 885– 891. http://www.informaworld.com/smpp/title~db=all~content= t713664979~tab=issueslist~branches=18 – v18
Bossink, B. A. G. (2002) "A Dutch public-private strategy for innovation in sustainable construction", *Construction Management and Economics* Vol. 20 No. 7, pp. 633–642.

Bressers, H. T. A. and De Bruijn, T. J. N. M. (2005) "Environmental voluntary agreements in the Dutch context", in Croci, E. (Ed.), *The Handbook of Environmental Voluntary Agreements, Design, Implementation and Evaluation* Springer, Dordrecht, pp. 261–281.

Brounen, D., *et al.* (2009) *Energy Performance Certification in the Housing Market Implementation and Valuation in the European Union*, available at http://www. corporate-engagement.com/index.php?pageID=1882&n=327 (accessed 10 December 2009).

DGBC (2009) "BREEAM-NL is een feit" (BREEAM-NL is a fact), available at http:// www.dgbc.nl/mediaroom/actueel/breeam-nl_is_een_feit/ (accessed 10 December 2009).

EC (European Commission) (2008) "Proposal for a Directive of the European Parliament and of the Council on the energy performance of buildings (recast) presented by the Commission", Brussels, 13.11.2008 COM (2008) 780 final, available at http://ec.europa.eu/energy/efficiency/buildings/buildings_en.htm (accessed 3 November 2009).

ENDS Europe (2005) "Holland rejects EU buildings energy law", ENDS Europe Friday 30 September 2005, available at http://www.endseurope.com/11051?view (accessed 2 November 2009).

Energids (2006) "Controle epc niet waterdicht" (EPC control not watertight), available at http://www.energiegids.nl/praktijkinformatie/details.tiles?doc=/cont ...s=all&toPath=/content/energie/publicaties/nieuwsblad-stromen/2006/07 (accessed 2 November 2009).

Energy Transition Task Force (2006) *More with Energy: Opportunities for the Netherlands*, available at http://www.senternovem.nl/energytransition/downloads/ index.asp (accessed 27 April 2009).

Gauzin-Müller, D. and Favet, N. (2002) *Sustainable Architecture and Urbanism: Concepts, Technologies, Examples*, Basel, London.

Glasbergen, P. (1998) "Partnership as a learning process. Environmental covenants in the Netherlands", in Glasbergen P. (Ed.), *Co-operative Environmental Governance. Public-Private Agreements as a Policy Strategy*, Dordrecht: Kluwer Academic Publishers.

Glasbergen, P. (1999) "Tailor-made environmental governance: on the relevance of the covenanting process", *European Environment* Vol. 9 No. 2, pp. 49–58.

Golub, J. (1998) *New Instruments for Environmental Policy in the EU*, London: Routledge.

IEA (International Energy Agency) (2009) *Energy Policies of IEA Countries – Netherlands – 2008 Review*, OECD/IEA, Paris.

Itard, L. and Meijer, F. (2008) *Towards a Sustainable Northern European Housing Stock. Figures, Facts and Future*, Amsterdam: IOS Press.

Itard, L., *et al.* (2009) *Consumentenonderzoek Lenteakkoord* (Consumer Research-Spring Agreement), Research commissioned by the NVB Association for developers and building contractors on behalf of Spring Agreement covenant partners, OTB Research Institute, Delft.

Joosen, S. (2007) *Evaluation of the Dutch Energy Performance Standard in the Residential and Services Sector within the Framework of the AID-EE Project*, available at http://www.aid-ee.org/documents.htm (accessed 10 December 2009).

Kuindersma, P. and Ruiter, C. J. W. (2007) *Eindrapportage Woonkwaliteit Binnenmilieu in Nieuwbouwwoningen, Eindresultaten van: 78 projecten/154*

Woningen (Final Report Indoor Climate Quality in New Build: final results of 78 projects/154 houses). Conducted by Adviesburo Nieman B.V. for VROM-Inspectie Regio Oost. Article code: 7559, available at http://www.vrominspectie. nl/actueel/publicaties/eindrapporta-gewoon-kwaliteit-binnenmilieu-in-nieuw bouwwoningen.aspx (accessed 10 December 2009).

Liefferink, D. (1998) "New environmental policy instruments in the Netherlands", in Golub, J. (Ed.), *New Instruments for Environmental Policy in the EU*, London & New York: Routledge.

Liefferink, D. and Van der Zouwen, M. (2003) "The Europeanisation of Dutch environmental policy: the advantages of being 'Mr. Average'?", In *European Union Studies Association (EUSA), Biennial Conference 2003 (8th), March 27–29, 2003, Nashville, Tennessee,* available at http://aei.pitt.edu/6518/ (accessed 10 December 2009).

Loorbach, D., *et al.* (2008) "Governance in the energy transition: practice of transition management in the Netherlands", *International Journal of Environmental Technology and Management*, Vol. 9, No. 2/3, pp. 294–315.

McKinsey & Company (2009) *Pathways to a Low Carbon Economy. Version 2 of the Global Greenhouse Gas Abatement Curve*, available at https://solutions. mckinsey.com/ClimateDesk/default.aspx (accessed 2 November 2009).

Melchert, L. (2007) "The Dutch sustainable building policy: a model for developing countries?" *Building and Environment* Vol. 42 No. 2, pp. 893–901.

Ministry for Economic Affairs (2007) *The Netherlands Energy Efficiency Action Plan 2007*, available at http://ec.europa.eu/energy/efficiency/end-use_en.htm (accessed 2 November 2009).

Moss, T., *et al.* (2005) "The politics of design in cities: preconceptions, frameworks and trajectories of sustainable building", in Guy, S. and Moore, S. (Eds.), *Sustainable Architectures: Cultures and Natures in Europe and North America*, Taylor and Francis e-Library pp. 73–88.

Opstelten, I. J., *et al.* (2007) "Bringing an energy neutral built environment in the Netherlands under control", in Seppänen, O. and Säteri, J. (Eds.), *Proceedings of Clima 2007 Wellbeing Indoors, Helsinki, 10–14 June 2007*, Finvac ry, Helsinki.

Ouroussoff, N. (2007) "Why are they greener than we are?" *New York Times* May 20 2007, available at http://www.nytimes.com/2007/05/20/magazine/20europe-t.html (accessed 10 December 2009).

Pettenger, M. (2007) "The Netherlands' climate policy: constructing themselves/constructing climate change", Paper presented at the annual meeting of the International Studies Association 48th Annual Convention, 28 February 2007, Chicago, available at http://www.allacademic.com/meta/p179125_index.html, (accessed 10 December 2009).

Priemus, H. (2005) "How to make housing sustainable? The Dutch experience", *Environment and Planning B: Planning and Design*, Vol. 32 No. 1, pp. 5–19.

Rovers, R. (2008) *Sustainable Housing Projects: Implementing a Conceptual Approach*, Amsterdam: Techne Press.

Schneider, H. and Jharap, R. (2010) "Signed, Sealed, Delivered? Evaluatie van drie convenanten energiebesparing in de gebouwde omgeving: Meer met Minder, Lente-Akkoord, Energiebesparing Corporatiesector, Eindrapport", (Evaluation of three covenants for energy saving in the built environment: More with Less, Spring Agreement and Energy Saving in the Corporation Sector, Final Report), available at http://www.lente-akkoord.nl/wat/publicaties/ (accessed 5 October 2010).

Van Bueren, E. (2009) *Greening Governance: An Evolutionary Approach to Policy Making for a Sustainable Built Environment*, Amsterdam: IOS Press BV.

Van Cruchten, G., *et al.* (2008) *Data Collection from Energy Audits for Residential Buildings of the Housing Corporations TBV Wonen and Wonen Breburg in Tilburg, Summary Report*, Arnhem: Builddesk Benelux BV.

Van den Brand, G. (2006) "Mapping tools for a sustainable building cycle", paper presented at PLEA 2006-The 23rd Conference on Passive and Low Energy Architecture, 6–8 September 2006, Geneva, Switzerland, available at http://www.unige.ch/cuepe/html/plea2006/proceedings.php (accessed 10 December 2009).

Van der Heijden, J. (2009) *Building Regulatory Enforcement Regimes: Comparative Analysis of Private Sector Involvement in the Enforcement of Public Building Regulations*, Amsterdam: IOS Press BV.

Van der Heijden, J. J., *et al.* (2007) "Problems in enforcing Dutch building regulations", *Structural Survey*, Vol. 25 No.3/4, pp. 319–329.

Van der Waals, J. and Glasbergen, P. (2002) "Policy innovations in the urban context", in Driessen, P.P.J. and Glasbergen, P. (Eds.), *Greening Society, the Paradigm Shift in Dutch Environmental Politics*, Dordrecht: Kluwer Academic Publishers.

Van Ekerschot, F. and Heinemans, M. (2008) "Implementation of the EPBD in the Netherlands: Status and Planning in June 2008", in Wouters, P. *et al.* (Eds.), *Implementation of the Energy Performance of Buildings Directive: Country Reports 2008*, Brussels: European Communities.

Van Kasteren, J., *et al.* (2002) *Buildings that Last: Guidelines for Strategic Thinking*, NAi Publishers in association with VROM and the Government Buildings Agency, The Hague.

Vereniging Eigen Huis (Dutch Home Owners Association) (2007) "Brief over uitstellen energielabel" (Letter regarding the postponement of the energy label), available at http://www.eigenhuis.nl/VerenigingEigenHuis/OverOns/Brieven+van+de+vereniging/BriefEnergielabelUitstellen.htm (accessed 4 November 2009).

Visscher, H. J. and Meijer, F. M. (2006) "Building regulations for housing quality in Europe", in Cernic Mali, B. *et al.* (Eds.), *ENHR Conference 2006: Housing in an Expanding Europe. Theory, Policy, Implementation and Participation, Urban Planning Institute of the Republic of Slovenia, Ljubljana*, pp. 1–9.

Visscher, H. J. and Meijer, F. M. (2008) "The growing importance of an accurate system of building control", in Carter, K. *et al.* (Eds.), *COBRA 2008. RICS Construction and Building Research Conference*, Royal Institution of Chartered Surveyors, London, pp. 1–14.

VROM (Ministerie van Volkshuisvesting, Ruimtelijke Ordening en Milieubeheer-Ministry of Housing, Spatial Planning and Environment) (2001) *Where there's a will there is a world. Fourth National Environmental Policy Plan*, available at http://www2.vrom.nl/Docs/internationaal/NMP4wwwengels.pdf (accessed 6 March 2009).

VROM (2007) *Nieuwe Energie voor het Klimaat: Werkprogramma Schoon en Zuinig* (New Energy for the Climate: Clean and Efficient Work Programme), available at http://www.vrom.nl/pagina.html?id=32950 (accessed 2 November 2009).

VROM (2009) "Duurzaam bouwen en verbouwen" (Sustainable Building and Renovation), available at http://vrom.nl/pagina.html?id=23963 (accessed 31 October 2009).

Winsemius, P. (1993) "Environmental contracts and covenants: new instruments for a realistic environmental policy?" in Van Dunne, J. M. (Ed.) *Environmental Contracts and Covenants: New Instruments for a Realistic Environmental Policy? Proceedings of an International Conference, Kononklijke Vermande, Lelystad.*

Zito, A. R. *et al.* (2003) "Instrument innovation in an environmental lead state: 'new' environmental policy instruments in the Netherlands", *Environmental Politics,* Vol. 12 No. 1, pp. 157–178.

4 The quest for sustainable buildings: getting it right at the planning stage

Julie Adshead

4.1 Introduction

Climate change and security of energy supply are key drivers of policy and legislation in current times. At the same time, the UK government is also focusing upon those households subject to 'fuel poverty'.[1] There is an impressive number of initiatives in place to secure reduction in greenhouse gas emissions and to promote energy efficiency. These measures range from legally binding international agreements to local voluntary community schemes. The complexity of the policy and legal frameworks is accentuated by the fact that there are multiple goals to achieve and this results in some incidences of paradox. Take micro generation, for example. Certainly domestic schemes will result in reduced carbon dioxide emissions and should provide a reliable source of energy for the future. As evidenced by the Climate Change and Sustainable Energy Act 2006, the UK government also sees micro generation as one of the hopes for reducing fuel poverty. Micro generation does, however, tend to be expensive (certainly in comparison with current energy prices) and if it is to be successful in the alleviation of 'fuel poverty' then substantial financial support will be needed. Ultimately, whether the economics of micro generation make sense will depend upon whether the era of cheap centralized energy is really at an end (Dow, 2007). The array of measures in place to improve energy performance and thus reduce carbon emissions will also serve to ease the burden of energy expenses on poor households. However, their role in addressing the goal of reducing carbon emissions is based on the premise that a large proportion of energy provision is from fossil fuel sources. A switch to a mix of nuclear and renewable sources (although this may raise other entirely different issues in relation to the environment and sustainability) would arguably be far more efficient in reducing carbon emissions from buildings.

This chapter focuses upon the legal provisions in place to reduce carbon emissions from buildings in the UK. In particular, it reviews the role and the potential of the UK planning regime to this end. The first section sets the context in terms of global and regional commitments to counter climate change. Some of the legislative provisions in the UK are then explored. This includes overarching measures such as the Climate Change Act, 2008 as well as specific legislation that targets the energy performance of buildings. This second section of the chapter also gives some brief attention to the non-legally binding 'Code for Sustainable Buildings' as this is integral to the way in which planning law can operate to improve standards. The third and final part of the chapter considers the role of the UK planning regime, including the development and future of the 'Merton Rule', incentives for micro generation through the 'Permitted Development Order' route and the use of model planning conditions. In order to illustrate how the use of planning conditions can succeed (or not), two case studies are considered. In one of the case studies, a planning condition was upheld whereas the relevant condition in the other case study was subject to a successful appeal. Some tentative conclusions are then drawn as to the future direction of law and policy in the UK relating to emissions from buildings in the context of the coalition government's 'localism' agenda.

4.2 The international and regional context

4.2.1 The international regime

The international climate change regime comprises the United Nations Framework Convention on Climate Change[2] and its Kyoto Protocol.[3] Both of these instruments are in force and are legally binding. The Kyoto Protocol commits many of the industrial nations to a reduction in the annual average of greenhouse gas emissions. In the 'first commitment period' from 2008 to 2012 the reduction is to an average level of 95 per cent of 1990 emissions. Fundamental problems have been identified with the Kyoto Protocol. Not least of these is the fact that the United States (US), responsible for 20 per cent of the overall output of greenhouse gases, is not a party (UNDP Human Development Report, 2007/2008). It has also been suggested that the commitments made so far are inadequate and have not been successfully implemented (Barker *et al.*, 2007, den Elzen, 2008). With the first commitment period drawing to a close in 2012, it was hoped that, following the Bali Action Plan, adopted by the international community in December, 2007, an 'agreed outcome' on long-term cooperative action on climate change would be reached in Copenhagen in December 2009. Despite much debate as to the possible legal form of the 'agreed outcome' (Rajamani, 2009) and high hopes of a legally binding agreement (Thomas and Woodward, 2010) there was no such agreement. The result of the Copenhagen Conference is an accord, led by the US, between China, India, Brazil, South Africa and the US to tackle global warming and deliver aid to developing nations. Despite criticisms of the Kyoto Protocol and the failure to reach any kind of multi-lateral agreement at Copenhagen, there is no doubt that the Kyoto goals have been a powerful driver for governments. The Copenhagen accord does provide for nations to commit to implement emissions targets for 2020 and a number of world leaders have signified their intention to introduce further more stringent, legally binding targets (Thomas and Woodward, 2010).

4.2.2 Law and policy of the European Union

The European Union (EU) has the quota of a reduction to 92 per cent of 1990 levels by 2012 under the Kyoto agreement. However, the EU is committed to even more stringent targets than those provided for under the international regime. The EU objective is to reduce overall greenhouse gas emissions by at least 20 per cent below 1990 levels by 2020 and by 30 per cent in the event of an international agreement being reached. The Union has also set a binding target for energy from renewable sources of 20 per cent of total EU energy consumption by 2020. A range of measures exist to achieve these goals, including a directive on energy end-use efficiency and energy services,[4] with an overall objective of saving 9 per cent of energy by 2012 and a directive on the promotion of the use of energy from renewable sources[5] that provides for the improvement of energy efficiency in the context of the binding target for energy from renewable sources. The key legislative instrument applying to the control of emissions from buildings is the directive on the energy performance of buildings (2002/91).[6]

Under directive 2002/91, member states are required to establish a methodology for determining the energy performance of buildings and set minimum energy performance standards. New buildings above 1000 square metres are to meet these standards as are buildings above this limit that undergo major renovation. On construction, sale or rent, an energy performance certificate for the building must be made available and this is to be no more than ten years old. Public buildings exceeding the 1000 square metres limit are required to display their certificates. The directive also provides for inspection regimes for boilers and air conditioning systems and for inspections to be carried out by independent experts (Hookins and Stonehill, 2006).

Over recent years there have been calls from the European Council and the European Parliament[7] for the Commission's priorities, established in its *Action Plan for Energy Efficiency: Realising the Potential* published in 2006,[8] to be comprehensively and swiftly implemented. The action plan identified the significant potential for cost-effective energy savings in the buildings sector and, as part of the package of measures to achieve the priorities of the action plan, a new directive on the energy performance of buildings was published in May 2010. When in force (2012), the directive will expand upon the provisions of the 2002 directive significantly (Mittenthal, 2009).

Under Article 4 of the new directive, member states will have to set minimum energy performance requirements for buildings or building units 'with a view to achieving cost-optimal levels'. This requirement will also apply to 'building elements that form part of the building envelope and that have a significant impact on the energy performance of the building envelope when they are replaced or retrofitted'. The requirement to meet minimum energy performance requirements will apply to all new and existing buildings, regardless of size.[9] Member states will also be required to set energy performance for technical building systems installed in existing buildings.[10] The directive includes a new binding requirement upon member states to ensure that by 2020 all new buildings are nearly zero-energy.[11] The display requirements for public buildings are extended to include those 'frequently visited by the public' and the threshold size is lowered to 500 square metres (to be reduced further to 250 square metres in 2015).[12] In addition, inspection requirements are extended to apply to all elements of heating systems (not just boilers).[13]

4.3 UK legislative provisions

4.3.1 The Climate Change Act 2008

The UK is also committed to more ambitious targets than those set in the Kyoto Protocol. In fact, it is the first nation worldwide to adopt a legally binding long-term framework to cut carbon emissions. The controversial Climate Change Act, 2008 imposes a statutory duty upon the Secretary of State of 2050 'to ensure that the net UK carbon account for the year 2050 is at least 80 per cent lower than the 1990 baseline'.[14] The imposition of statutory duties on government is a novel approach in UK law (Stallworthy, 2010) and some might doubt the meaningfulness, in particular, of imposing a legal duty on an individual whose identity is, as yet, unknown. It is also difficult to see how legally binding this target can be, when it is unlikely to be legally enforceable (Townsend, 2009; Stallworthy, 2010). However, proponents of the Act (Grekos, 2009; Townsend, 2009) recognize its possibilities in terms of improving carbon management, moving the UK towards a low carbon economy and providing strong leadership and commitment to shouldering an equitable burden in reducing global emissions. It may also provide some certainty and encouragement for industry and business.

The Act requires that a series of five-yearly carbon budgets are set by order of the Secretary of State. The first three budgets (2008–22) have already been set by the Carbon Budgets Order[15] with a view to meeting the 2050 target. Thus 2018–2022 is 34 per cent lower than the 1990 baseline. There is a more ambitious figure of 42 per cent by 2020, which will only be adopted if a global agreement is reached. The crediting of carbon units is going to be crucial if these objectives are to be met. This is clearly illustrated by the scale of net reduction from 2007 to 2008, which was a mere 2 per cent (DECC, 2010). The Carbon Accounting Regulations, which define carbon units and set out how carbon can be credited to the account, came into force on 31 May 2009.[16] Greenhouse gas allowances (under trading schemes) can also act as credits, but only credits under the EU emissions trading scheme can be credited to the UK carbon account.

The Secretary of State is subject to a duty 'to ensure that the net UK carbon account for a budgetary period does not exceed the carbon budget'.[17] Some indication as to how this will be achieved is provided in measures recommended by the Climate Change Committee set up under the Act to advise the Secretary of State.[18] The key short-term recommendations of the Committee are:

- Energy efficiency improvements in building and industry
- Fuel efficiency improvement in road vehicles
- A significant shift towards renewable and nuclear power generation and renewable heat

In order to achieve the extremely ambitious targets provided for in the 2008 Act, a range of policy and legislative initiatives have been put in place. The two primary mechanisms for the delivery of energy efficiency improvements in buildings are building regulations and the planning regime and these are considered below.

4.3.2 Powers under the Building Act 1984

The Building Act 1984 places certain aspects of building under statutory control and empowers the Secretary of State to make regulations that provide details of exactly how that control is exercised.[19] The scope of the Building Act in terms of regulating the conservation of fuel and power was significantly widened by two recent pieces of legislation: The Sustainable and Secure Buildings Act 2004 and the Climate Change and Sustainable Energy Act 2006. The Sustainable and Secure Buildings Act 2004 enables the making of regulations for 'furthering the conservation of fuel and power'[20] and extends the range of matters in respect of which regulations can be made.[21] Significantly it allows for the regulation of existing buildings in matters relating to energy conservation and carbon emissions.[22] It also inserts a new Section 2A into the Building Act 1984, which allows for regulations to be made that impose 'continuing requirements' on building owners and occupiers regardless of when the building was erected or whether other building works are ongoing. These powers are potentially far-reaching and would permit, for example, the making of regulations requiring all lofts to be insulated (McAdam, 2007). The Climate Change and Sustainable Energy Act 2006 further extended the powers under the Building Act 1984 by enabling regulations to be made under the Act relating to the installation of micro generation technologies in buildings.[23] The Act also extends the time limit for prosecution of those in breach of regulations specifically relating to the conservation of fuel and power.[24]

The Building Act 1984 allows for guidance documents to be approved and compliance with approved guidance creates a presumption that the works in question comply with the requirements of the Act as provided for in The

Building Regulations. Part L guidance documents deal with the conservation of fuel and power[25]and these cover new dwellings, existing dwellings and new and existing buildings other than dwellings. Under The Building Regulations, target emission rates for buildings have to be set for new dwellings, and buildings over 1000 square metres must be brought up to Part L standard when renovated.[26] For new dwellings the government has committed to a programme by which regulations will demand 25 per cent lower carbon emissions by 2010, 44 per cent lower by 2013 and by 2016 all new dwellings should be zero carbon.[27] Thus, the Part L standard in building regulations will be incrementally raised over forthcoming years. New Part L guidance implementing the 25 per cent lower emissions requirement came into force in October 2010. For existing housing stock, it is arguable that the full potential of the Building Act and regulations made under it have not as yet been fully realized. The government has turned its attention to the problem of tackling emissions from existing housing and a House of Commons Report was published on the subject in 2008.[28] The report sets out recommendations for improving energy efficiency in existing housing with a 'shopping list' of recommended measures for government (Grekos, 2008). These include measures to encourage the take-up of home micro generation, requirements for consequential energy efficiency improvements in planning consent on extension of homes, new requirements for Energy Performance Certificates and the production of a 'Code for Existing Homes' along the lines of the 'Code for Sustainable Homes' (see below).

4.4 The Code for Sustainable Homes

The Sustainable Buildings Task Group first proposed a Code for Sustainable Buildings in 2004. The idea was that the voluntary code would be a catalyst for low-carbon, low-impact building and set vanguard eco-standards for the government to follow. The Code for Sustainable Homes was finally launched in December 2006 and the technical guide followed in April 2007. The code sets six standards of increasing rigour against which the whole home can be measured. A whole range of factors are considered alongside carbon dioxide emissions. There are nine design categories and the levels are rated from one to six stars. Level 6 (six stars) of the code, in terms of emissions, is 'true zero carbon'. The lowest, one-star level was, until 2010, more demanding than minimum standards for building regulations. However, new Part L standards came into force in October 2010 which equate to level three of the Code. The Part L standard for emissions from buildings is now equivalent to level 3 (three stars) of the code.

At its inception, the code was a voluntary mechanism, but since May 2008, sellers of new properties have been required to provide information to the purchaser on the rating of the building, either in the form of a code certificate or a statement of non-assessment. Also from 2008, achievement of code level 3 became mandatory for all publicly supported developments.

The first proposals for a code were intended to embrace both new homes and non-domestic buildings and calls continue for the code to be expanded in this respect (UKGBC, 2009) as well as for a similar code to be adopted for existing homes (see above). Although essentially still a voluntary system, as shall be demonstrated below, when used in tandem with the UK planning system it has the potential to drive up standards across the entirety of the nine design categories, and thus serve to increase the chances of reaching the overarching emission targets set by government in the Climate Change Act.

4.5 The UK planning regime

A full account of the UK planning regime is beyond the remit of this chapter, but an overview of certain elements of UK planning law is helpful in understanding how the planning system may operate to drive up emissions performance.

4.5.1 Development

The UK system is centred upon 'development'. The key statute, the Town and Country Planning Act 1990, states that 'Planning permission is required for the carrying out of any development of land'.[29] The definition of 'development' in the Act is extremely broad and encompasses both building operations and the change of use of buildings.

> Development ... means the carrying out of building, engineering, mining or other operations in, on, over or under land, or the making of any material change in the use of any buildings or other land.[30]

4.5.2 The development plan

UK planning is led by the development plan and reference to the plan will be the starting point in determining a planning application. Development plans are broad, giving general policy, aims, objectives and goals and are generally permissive in nature. The Planning and Compulsory Purchase Act 2004 introduced a new range of strategies and plans. The 'Regional Spatial Strategy' introduced under the 2004 Act has since been suspended by the current government, but the local development framework (including the development plan) continues to operate. The Planning and Compulsory Purchase Act states that

> regard is to be had to the development plan, the determination shall be made in accordance with the development plan unless material considerations indicate otherwise.[31]

4.5.3 Material considerations

Although the development plan provides the starting point, it will not necessarily be the dominant determinant in the decision.[32] Material considerations can be taken into account and on occasion they can win out over the development plan.[33] There is no statutory definition or guidance as to what constitutes a material consideration. Certainly planning guidance, representations, as well as many other matters determined by case law, will be material considerations. Environmental considerations are just one of many possible elements to be brought into the balancing act in the determination of a planning application.

4.5.4 Planning guidance

Planning guidance plays a pivotal role in the UK planning system. There are a range of guidance documents in place, which, as noted above, will constitute material considerations in the determination of a planning application. The statutory requirement in the Planning and Compulsory Purchase Act 2004 for all plan making bodies to exercise their functions 'with the objective of contributing to the achievement of sustainable development' is reflected in the government's Planning Policy Statement (PPS)1: Delivering Sustainable Development and its supplement, Planning Policy Statement: Planning and Climate Change as well as Planning Policy Statement 22: Renewable Energy. A new draft climate change planning policy statement was released in 2010,[34] which combines the policies currently set out in the supplement to PPS1 and PPS22.

4.5.5 Planning conditions

Almost all planning determinations include conditions. The Town and Country Planning Act, 1990 allows the local planning authority to impose such conditions 'as it thinks fit'.[35] Guidance is given in the act[36] and in the Secretary of State's policy on conditions.[37] The courts have taken quite a restrictive view on planning conditions and, in the case of *Newbury District Council v Secretary of State for the Environment*,[38] it was held that conditions must

- be imposed for a planning purpose and not for an ulterior motive
- fairly and reasonably relate to the development permitted
- not be perverse (so unreasonable that no reasonable authority could have imposed them)

4.5.6 General Permitted Development Order

There are certain ways in which development can be permitted under statute and thus the need for a planning application is obviated. One such route is available under the General Permitted Development Order.[39] There are a large number of development types listed in Schedule 2 that cover minor developments, developments carried out by public services and favoured activities (such as agriculture and forestry). Development consent may still be needed when projects exceed certain thresholds and the right to development can be withdrawn, for example, if an environmental impact assessment is required.

4.6 Planning and micro generation

The 2007 Government White Paper 'Meeting the Energy Challenge'[40] recognized that planning consents are a major constraint on the implementation of a future energy strategy. Indeed the Scottish Executive has acknowledged that the planning system has the potential to act as a tool in opposing nuclear power (Dow, 2007). There is certainly little doubt that barriers have been encountered in the development of micro generation, but conversely the planning system has also been a driver behind the adoption of on-site renewable energy, which will have a significant role to play in delivering the government's 2016 zero carbon homes agenda (Sustainable Energy Partnership, 2007).

4.6.1 The Merton Rule

The Merton Rule takes its name from Merton Council. It amounts to a borough-wide prescriptive planning policy for all buildings, which was developed and adopted by the council in 2003. The policy requires new developments to generate at least ten per cent of energy needs from on-site renewable technology. The normal threshold for application of the rule is ten homes or 1,000 square metres of non-residential development. The Merton policy has had a significant impact and was subsequently adopted by the Mayor of London and the majority of local authorities nationwide. Planning Policy Statement 22 on renewable energy expressly acknowledges the Merton Rule and advocates its adoption by local planning authorities and the encouragement of renewable energy projects through local planning documents.

However, the future of the rule is uncertain. Whilst Merton Council intends to extend the policy to cover all development in Merton and is considering whether it is appropriate to increase the percentage of the policy up to a 20 per cent requirement (Merton Council, 2010), it has been suggested (Sustainable Energy Partnership, 2007) that the rule was watered down in the 2007 Climate Change supplement to PPS1 by the removal of the requirement to consider renewable energy projects 'in **all** new developments'. The emphasis also seems to have shifted from a requirement for a percentage of on-site renewable energy to consideration of the possibility of utilizing off-site renewable energy supplies. This latter development is subject to criticism as, it is suggested that, linking new housing to off-site renewable energy developments provides no additional cut to carbon dioxide emissions (Sustainable Energy Partnership, 2007). The recent draft supplement to PPS1 essentially prohibits the adoption of the Merton Rule in the future, stating that

> targets for application across a whole local authority area which are designed to secure a minimum level of decentralised energy use in new development will be unnecessary when the proposed 2013 revisions to Part L of the Building Regulations . . . are implemented.

However, it should be noted that currently a disclaimer appears on the Department for Communities and Local Government website, where the draft PPS appears, warning that all the content is subject to review in the light of the recent change of government.

4.6.2 Permitted Development Order

One of the problems with micro generation and the planning system traditionally lies with local authorities taking different approaches to small household projects. For example, some allow small turbines on houses whereas some do not and similarly with solar panels, whilst allowed in some areas, other local authorities view panels as damaging to conservation areas (Dow, 2007). The Climate Change and Sustainable Energy Act 2006 allows for such differences in practice to be reduced through amendment to the Permitted Development Order. After due consultation, the Secretary of State made an order[41] amending the General Permitted Development Order[42] allowing permitted development for the installation of domestic micro generation equipment. Within the framework of the restrictions and conditions outlined in the order, solar panels, heat pumps and biomass heating systems are all subject to the order. This should make it easier for households to install micro generation equipment because, in many cases, it will no longer be necessary to apply for planning permission. It should also lead to greater consistency across local authority areas.

4.7 Planning conditions and sustainable buildings

The supplement of Planning Policy Statement 1 on climate change urges planning authorities and developers to 'engage constructively and imaginatively to encourage the delivery of sustainable buildings'. The statement also acknowledges that 'There will be situations where it could be appropriate for planning authorities to anticipate levels of building sustainability in advance of those set out nationally'. Planning authorities are advised to focus on development area or site-specific opportunities and specify requirements in terms of nationally recognized standards such as the Code for

Sustainable Homes. Some local planning authorities have sought to achieve the goal of driving up standards of sustainability in new domestic dwellings in their areas by recommending enhanced levels of performance and certification under recognized codes and schemes.

4.7.1 Brighton and Hove City Council

An example of this practice is to be found in the 'Model Planning Conditions and Informatives' of Brighton and Hove City Council, adopted prior to the changes to Part L standards in October 2010 (see above). The model planning conditions state that, unless otherwise agreed in writing by the Local Planning Authority, no new-build residential development can commence without evidence that it will achieve a minimum code level 3 under the Code for Sustainable Homes. Furthermore (again, unless agreed in writing), the approved units cannot be occupied until a final code certificate is issued confirming the minimum code level 3 performance. Similarly, the model conditions state that unless agreed with the Local Planning Authority, no residential development involving existing buildings can commence unless it is certified that the development will achieve an 'Ecohomes' rating and no occupation can take place until a post construction certificate to this effect has been submitted to and approved by the authority. In a similar way, BREEAM registration, assessment, rating and confirmation are conditions of development and occupation for new build non-residential developments. Such conditions in planning determinations might, however, be open to challenge as is illustrated by the two following case studies.

4.7.2 The former New Penny public house

Planning permission had been granted subject to conditions for the construction of twelve new-build flats on the site of a former public house. An appeal was brought by the developers against the decision of Cheltenham Borough Council. One single condition to the planning permission was disputed. The condition in question stated that

> prior to the commencement of development a scheme to demonstrate a reduction in carbon emissions to achieve a minimum level of code 3 of the Code for Sustainable Homes shall be submitted to and approved in writing by the local planning authority. The development shall be carried out and maintained in accordance with the details so approved.

The reasons given for the condition were to ensure compliance with national and regional objectives and the aims of local plan policy CP1 regarding sustainable development. The latter policy stated that development would be permitted only where it took adequate account of the principles of sustainable development and it set out a number of criteria for this. The policy also stated a number of principles of sustainable development which might be taken into account as material considerations in the determination of planning applications.

The appeal was allowed, the inspector concluding that the condition was imprecise and unreasonable and that it did not meet the tests set out in Circular 11/95. A new planning permission was granted without the disputed condition but retaining the relevant non-disputed conditions from the previous permission. The key reasons for allowing the appeal were as follows:

- The Supplement to PPS1 (the Supplement), states that councils wishing to proceed in advance of the Government's timetable must set out their policies for sustainable developments in development planning documents (DPDs) so as to ensure examination by an independent inspector. Neither policy CP1, *Sustainable Development of the Cheltenham Local Plan Second Review* nor the Council's supplementary planning guidance documents relating to policy CP1 referred to the Code for Sustainable Homes.
- Paragraph 42 of the Supplement allows non-compliance with adopted DPD policies if this is not feasible or viable. The appellant argued that the development had not been designed to attain code level 3 and that fundamental changes, involving a different design to that which had been approved, would be required to achieve this. The cost involved, it was argued, would make the proposed development unviable. Little evidence was provided by the council to counter these claims.
- The approach set out in the Supplement indicates that conditions requiring compliance with level 3 of the code or above should only be imposed if the developer has demonstrated a willingness to comply. Condition 13 was unreasonable without the developer's agreement to it.
- The developer had committed to construction of the flats in accordance with the Code to a level to be determined at the time of construction and to a number of measures of sustainable methods of construction. The Inspector concluded that the measures incorporated in the scheme met the sustainability objectives identified in policy CP1. It was also noted that it was normal to assess the code level of a development post-completion of the development rather than pre-commencement as required by condition 13.

(Appeal Decision, 2009)

4.7.3 Hut Cottage

Planning permission had been granted subject to conditions for a replacement bungalow at Hut Cottage, a single-storey dwelling with outbuilding on a small and irregularly shaped plot of land abutting listed buildings and in a conservation area. Planning permission had been granted for a replacement bungalow in 2005, which had expired. A further application of 2008 seeking renewal of the planning permission was for a proposed development identical to the earlier scheme. When planning permission was granted it was subject to conditions, one of which was disputed in an appeal against the decision of Chelmsford Borough Council. The condition in dispute (No. 5) stated that, unless otherwise agreed in writing by the Local Planning Authority:

a) The development hereby permitted shall be built to a minimum of level 3 of the Codes for Sustainable Homes (or its successor);
b) No development shall take place until a design stage assessment (under the Code for Sustainable Homes or its successor) has been carried out and a copy of the summary score sheet and Interim Code Certificate have been submitted to and approved in writing by the Local Planning Authority;
c) Prior to the first occupation of the dwelling, a copy of the summary score sheet and Post Construction Review Certificate (under the Code for Sustainable Homes or its successor) shall be submitted to the Local Planning Authority verifying that the agreed standards have been met.

The reasons given for the condition were to achieve sustainable development in accordance with Policies CP11 and DC24 of the Adopted Core Strategy and Development Control Policies Development Plan Document (CSDCP) and the Sustainable Development Supplementary Planning Document (SPD). Policy CP11 provided guidance relating to energy and resource efficiency, renewable energy and recycling and Policy DC24 established criteria relating to energy-efficient design and the use of materials, including a requirement that all new dwellings should attain a minimum rating of level 3 of the Code for Sustainable Homes or its successor.

The appeal was dismissed and the disputed condition was found to be reasonable and necessary. The reasons given by the inspector were as follows:

*   There had been changes in policy in response to growing concerns surrounding global warming since the first permission was granted, which were reflected in the 2007 Supplement to PPS1. The Council's determination and conditions were guided by the policies within the CSDCP.
*   It was noted that the Planning and Compulsory Purchase Act 2004[43] states that regard must be had to the development plan unless material considerations indicate otherwise.
*   It was also noted that where planning permission expires, fresh applications should be judged against current planning considerations.[44]
*   The decision to reject the appeal was made notwithstanding the resultant financial implications and the voluntary nature of the Code for Sustainable Homes.

(Appeal Decision, 2009)

4.8 The future of requirements for sustainable buildings in local planning

The two appeals outlined above deliver some interesting lessons if planning conditions are to be successfully utilized to attain higher standards than currently required under building regulations. Clearly, the requirement of attaining a particular level of an accepted national code such as the Code for Sustainable Homes is, in principle, acceptable (at least under current planning guidance). The key message to be drawn from the different outcomes of the two case studies above is that explicit reference to the code in question should be contained within the development plan document (DPD). The question as to whether requirement can be made for schemes of compliance to a level of performance pre-commencement rather than post-completion is less clear cut. To err on the side of caution local planning authorities might be best advised to stick to post-completion requirements in order to minimize the chances of successful appeal. It is interesting that in the Hut Cottage appeal, the fact that the condition was not imposed with the agreement of the developer was given little weight. Also, although the Inspector noted the financial consequences of the imposition of condition No. 5, the issues of the feasibility and viability of compliance with level 3 of the code were not really explored in depth.

Although the future of the new draft supplement to PPS1 is uncertain, the draft policy on the local planning approach to setting requirements for sustainable buildings, as it is currently framed, would continue to allow requirements for a building's sustainability as long as they are set out in the DPD. The approach is, however, considerably more restrictive than that in the current supplement. Requirements should

> relate to a development area or specific sites and not be applicable across a whole local authority area unless the justification for the requirement can be clearly shown to apply across the whole area.[45]

If this is retained in any new PPS, then it will be much harder for local authorities to adopt a requirement to meet elevated levels of a code across their whole area. Certainly they will have to provide clear and convincing justification within their DPD if this is to be the case.

4.9 Conclusions

The law surrounding climate change and energy efficiency is complex and multi-layered. This complexity is accentuated by the multiple aims involved and the fact that measures adopted do not always align to all of these. What is clear is that buildings are of central importance in realizing international and local goals to reduce levels of greenhouse gas emissions and thus help counter the threat of climate change. There are many ways of tackling the reduction of emissions from buildings. Alternative energy sources, renewables and energy efficiency are just some of these. Legislation will not provide the whole solution, nor will one single legislative route.

Regulations adopted under the Building Act 1984 are clearly of key importance in implementing the standards required of buildings in order to meet the targeted reductions in carbon dioxide emissions and the ultimate aim of carbon zero buildings. However, voluntary schemes, such as the Code for Sustainable Homes, also have an important role to play. This is clearly illustrated by the way in which the new Part L standards in the Building Regulations mirror level 3 of the code as well as the way in which the code has been used to drive up standards through the planning process.

New proposals for planning guidance appear to suggest that in some respects the role of the planning system is done. The proposal is to outlaw initiatives such as the 'Merton Rule' and restrict the requirements that can be imposed at the planning stage on the sustainability of buildings. Whilst there is doubtless some merit in having a single central level of control in the guise of building regulations, the building control system operates in a different way and at a different stage in the building life cycle to the planning regime. Having a single central standard also ignores variations in local environments and stifles the drive to strive for better standards and develop new and affordable technologies, which have in the past been encouraged by voluntary code ratings and compulsory on-site renewable energy requirements.

There is great uncertainty at the moment as to the direction that planning law will take. The new UK government promises radical reform of the planning system and has pledged through its proposed Decentralisation and Localism Bill[46] to devolve greater powers to local authorities. Specifically, it plans to abolish Regional Spatial Strategies and 'return decision-making powers on housing and planning to local councils'.[47] It may, therefore, be that the trend outlined above will be reversed and local communities and their planning authorities will be able to lead the way in delivering alternative energy sources and sustainable buildings.

Notes

1 Following the Warm Homes and Energy Conservation Act 2000, 'fuel poverty' is generally accepted as a consumer spend of more than 10 per cent per week of income on energy.
2 United Nations Framework Convention on Climate Change (adopted 29 May 1992, entered into force 21 March 1994) 1771 UNTS 107.
3 Kyoto Protocol to the United Nations Framework Convention on Climate Change (adopted 10 December 1997, entered into force 16 February 2005) 37 ILM 22.
4 Dir 2006/32 on energy end-use efficiency and energy services OJL 114 27.4.2006 p64–85.
5 Dir 2009/28 on the promotion of the use of energy from renewable sources OJL 140 5.6.2009 p16–62.
6 Dir 2002/91 on the energy performance of buildings OJL 1 4.1.2003 p65–71.

7 European Council of March, 2007, European Parliament Resolutions of 31 January 2008 and 3 February 2009.
8 Communication from the Commission, *Action Plan for Energy Efficiency: Realising the Potential*, Brussels, 19.10.2006 COM (2006) 545 final.
9 Dir 2002/91, Articles 6 and 7.
10 Dir 2002/91, Article 8.
11 Dir 2002/91, Article 9.
12 Dir 2002/91, Article 13.
13 Dir 2002/91, Article 14.
14 Climate Change Act, 2008, Section 1.
15 SI 2009, No. 1259.
16 SI 2009 No. 1257.
17 Climate Change Act 2008, Section 4 (1).
18 Building a low-carbon economy – the UK's contribution to tackling climate change published on 1st December 2008.
19 Building Act 1984, Sections 1, 2, 2A and Schedule 1.
20 Sustainable and Secure Buildings Act 2004, Section 1 (1) (b).
21 Building Act 1984, Section 1A as inserted by Sustainable and Secure Buildings Act 2004, Section 1 (3).
22 Sustainable and Secure Buildings Act 2004, Section 3 (7) (5).
23 Climate Change and Sustainable Energy Act 2006, Section 11.
24 Ibid., Section 13.
25 Building Regulations 2000, Schedule 1.
26 Implementing Directive 2002/91 (Number 6, above).
27 Meaning that, during the course of a year, the net carbon emissions from all energy use in the building is zero.
28 *Existing Housing and Climate Change*, House of Commons Communities and Local Government Committee, Seventh Report of Session 2007/8, 2, April 2008.
29 Town and Country Planning Act, 1990, Section 57 (1).
30 Ibid., Section 55 (1).
31 Planning and Compulsory Purchase Act 2004, Section 38 (6).
32 City of Edinburgh Council v Secretary of State for Scotland [1998] JPL 224; R v Leominster District Council ex parte Pothecary [1998] JPL 335.
33 R (on the application of the Council for National Parks Ltd.) v Pembrokeshire Coast NPA and ors. [2005] EWCA Civ 888.
34 Planning for a Low Carbon Future in a Changing Climate, 9 March 2010.
35 Town and Country Planning Act 1990, Section 70 (1).
36 Ibid., Sections 72 and 75.
37 Circular 11/95.
38 [1981] AC 578.
39 Town and Country Planning (General Permitted Development Order) 1995, SI 1995/418.
40 Meeting the Energy Challenge: A White Paper on Nuclear Power, Cm 7296, 10 January 2008.
41 The Town and Country Planning (General Permitted Development) (Amendment) (England) Order 2008.
42 Town and Country Planning (General Permitted Development) Order 1995.
43 Planning and Compulsory Purchase Act 2004, Section 38 (6).
44 Circular 08/2005, Guidance on Changes to the Development Control System.
45 PPS: Planning for a Low Carbon Future in a Changing Climate, Part 2; Consultation Draft, Policy LCF9.1 (i).
46 Queen's Speech, 25/5/10.
47 Department for Communities and Local Government Draft Structural Reform Plan, July 2010.

References

Appeal Decision (2009) 'Appeal Ref: APP/W1525/A/09/2096523 Hut Cottage, Colemans Lane, Danbury, Chelmsford CM3 4DN' by Roger P Brown. Decision date 17 June 2009.

Appeal Decision (2009) 'Appeal Ref: APP/B1605/a/09/2097989 Former New Penny Public House, 84 Gloucester Road, Cheltenham, Glos. GL51 8NZ' by Elaine Benson. Decision date 11 June 2009.

Barker, T. *et al.* (2007) 'Climate Change 2007: Mitigation of Climate Change: Contribution of Working Group III to the Fourth Assessment Report of the Intergovernmental Panel on Climate Change', Cambridge University Press, Cambridge, Box 13.7, p 776.

DECC (Department of Energy and Climate Change) (2010) 'Final Report 2008', http://www.decc.gov.uk/en/content/cms/statistics/climate_change/gg_emissions/uk_emissions/2008_final/2008_final.aspx, viewed 30.7.2010.

den Elzen, M. (2008) 'Emission Reduction Trade-Offs for Meeting Concentration Targets', Bonn Climate Change Talks, IPCC workshop, UNFCCC SBSTA 28.

Dow, S. (2007) 'Climate Change and Sustainable Energy Act 2006', *Environmental Law Review*, Volume 9 (4), pp 279–284.

Energy Saving Trust 'Code for Sustainable Homes: New Build Housing', Briefing Note, http://www.energysavingtrust.org.uk/housingbuildings, viewed 30.7.2010.

Grekos, M. (2008) 'House of Commons Report on Tackling Emissions from Existing Housing', *Journal of Planning and Environment Law* (8), pp 1123–1124.

Grekos, M. (2009) 'Climate Change Act 2008', *Journal of Planning and Environment Law*, Volume 4, pp 454–455.

Hookins, D. and Stonehill, C. (2006) 'Energy Performance of Buildings Directive – Time for an Energetic Approach', *Landlord and Tenant Review*, Volume 10 (6), pp 175–180.

McAdam, W.B. (2007), *Built Environment Carbon Emissions: A Policy Review*, MSc Construction Law and Arbitration Dissertation, Leeds Metropolitan University.

Merton Council, http://www.merton.gov.uk/living/planningpolicy/mertonrule, viewed 30/7/2010

Mittenthal, L. (2009) 'Update on Progress Towards Enactment of Revised Energy Performance of Buildings Directive', *Environmental Liability*, Volume 17 (5), pp 78–79.

Rajamani, L. (2009) 'Addressing the 'Post-Kyoto' Stress Disorder: Reflections on the Emerging Legal Architecture of the Climate Regime', *International and Comparative Law Quarterly*, Volume 58 (4), pp 803–834.

Stallworthy, M. (2010) 'UK Climate Change Law: Securing a Low Carbon Economy by Recourse to Experts and Government Obligations?', Proceedings of RICS COBRA Conference, Paris, 2nd & 3rd September, 2010.

Sustainable Energy Partnership (2007) 'The Merton Rule: The Orwellian World of Sir Humphrey, Minus Means Plus', http://www.microgenscotland.org.uk/SEP_Broadsheet_2%5B2%5Dfinal16.10.pdf, viewed 30.7.2010.

Thomas, M. and Woodward J. (2010) 'Copenhagen: Hopenhagen? Unfortunately not', *European Lawyer*, (92), p 7.

Townsend, H. (2009) 'The Climate Change Act 2008: Something to be Proud of After All?', *Journal of Planning and Environment Law* (7), pp 842–848.

UNDP (United Nations Development Programme) (2007/2008), 'Human Development Report, Fighting Climate Change: Human Solidarity in a Divided World', UNDP, http://hdr.undp.org/en/media/HDR_20072008_EN_Complete.pdf, viewed: 30.7.2010.

UKGBC (UK Green Building Council) 'Making the Case for a Code for Sustainable Buildings' March 2009.

5 Green buildings: a critical analysis of the Turkish legislation

Deniz Ilter

5.1 Introduction

Green buildings are defined as resource-efficient and ecosystem-conscious structures designed with a holistic understanding of social and environmental responsibility, harmonious with local conditions, built with proper materials and systems so as to minimize energy consumption, where renewable energy sources are given priority and waste production held under control (Turkish Green Building Association, 2009). The notion of "green building" pertains to the whole life cycle of a building from site selection to demolition. Also variously designated as "ecological," "environment-friendly," or "sustainable" buildings, green buildings have become one of the most important elements in the issue of sustainability, an issue inevitably highlighted by climate change and depletion of natural resources.

"Sustainable development" has been defined in a report by the United Nations–sponsored World Commission on Environment and Development (WCED) in 1987, as *"development that meets the needs of the present without compromising the ability of future generations to meet their own needs"* (WCED, 1987, p. 43). Unfortunately today, the construction, maintenance, and use of buildings transgress the principles of sustainable development and are currently contributing significantly to irreversible changes in the world's climate, atmosphere, and ecosystem (Akbiyikli *et al.*, 2009). Buildings are responsible for 30 percent of waste output, 39 percent of energy use, 38 percent of CO_2 emissions, 72 percent of electricity consumption, 40 percent of raw material use, and 14 percent of potable water consumption (US Green Building Council, 2009). With the aim of addressing this harm, the concept of "green buildings" has gained recognition in the architecture, engineering, and construction (AEC) industry within the past decade and today the AEC industry is facing ever-increasing demand to improve its sustainability performance.

This paradigm shift also generated the need to develop norms and regulations for sustainable building design and construction. Today there is a well-developed green building certification practice and a diverse array of regulations regarding green buildings and sustainability in several countries.

However, the results of research by Akbiyikli *et al.* (2009) show that the current level of sustainability understanding and hence its implementation is still unstructured, piecemeal, and insufficient in the Turkish AEC industry.

Besides the problems regarding energy and environment, with the joint effects of the Kyoto Protocol signed in February 2009 and the process of adopting related European Union statutes as an accession country, Turkey is on the verge of a revolution in "sustainability," and in "energy efficiency" – a major prerequisite of sustainability. Climate change, the situation of energy resources in Turkey and in the world, alternative sources of energy, and energy efficiency technologies are at the top of Turkey's agenda, together with green buildings and energy conservation in industry, transport, and buildings. The preparation of relevant legislation constitutes a decisive part of Turkey's transformation process. In the last couple of years, Turkey has enacted an array of new primary and subordinate legislation on sustainability and energy efficiency issues that are compliant with the pertinent European Union legislation; yet, currently no specific legislation exists on green buildings. The Energy Efficiency Law promulgated in 2007, regulations and by-laws put into force in connection with it in 2008 and 2009, such as the Regulation on Energy Performance of Buildings and the National Building Energy Performance Calculation Methodology, are still being discussed among academicians and practitioners in Turkey in this context.

Following a general overview in the next section, of economy, legal system, energy and environment in Turkey, this study analyzes the current legislation in the framework of "green building themes" developed on the basis of performance criteria used in the green building certification systems. The final section draws attention to the deficiencies in policies and statutory regulation pertaining to green buildings and brings forth proposals for the promotion of green buildings in Turkey.

5.2 Country overview

5.2.1 Economy

Turkey is a founding member of the Organisation for Economic Co-operation and Development (OECD) and the G-20 major economies. The gross domestic product (GDP) growth rate from 2002 to 2008 averaged 6 percent (Turkish Statistical Institute, 2009), which made Turkey one of the fastest growing economies in the world. The International Monetary Fund (IMF) forecasts a GDP[1] of 999 billion USD for Turkey in 2009 (International Monetary Fund, 2009) making her the fifteenth largest economy in the world.

The European Union (EU) accession process, particularly the Customs Union between Turkey and the EU Countries and the conclusion of the Uruguay Round are the main determinant factors shaping Turkey's international trade policies and orientations (World Trade Organization, 1998).

Key sectors of the Turkish economy are tourism, construction, banking, home appliances, electronics, textiles, oil refining, petrochemical products, food, mining, iron and steel, machine industry, and automotive.

The construction industry represents 6 percent of Turkey's GDP by itself; together with all complementary and related industries, it accounts for 30 percent of the GDP (Turkish Contractors Association, 2009). The industry is divided into two parts: the lower-quality domestic-only set of firms and the higher-quality international firms (Katsarakis *et al.*, 2007). On the lower-quality side, there are more than 30,000 active local firms, whereas 145 members of the Turkish Construction Association stand on the international side, 31 of which are among the 225 world's largest construction companies in terms of overseas activities (Engineering News-Record, 2009). From 1972 to 2008, Turkish Contractors have undertaken almost 5,000 projects in 70 countries, with a combined value of 130 billion USD (Turkish Contractors Association, 2009). Currently, Turkish firms have a very strong presence in the Middle East, North Africa, Former Soviet Union, and Eastern Europe, and have started penetrating Western Europe.

This level of activity, particularly the international activity, places upon the Turkish construction cluster a responsibility to play a pivotal role in sustainable construction by effective protection of the environment and careful use of natural resources (Akbiyikli *et al.*, 2009).

5.2.2 Legal system

The Turkish Law and legal system, compared to the Anglo-American Common Law, falls under the Civil Law tradition of continental Europe, based on statutory and legal enactments. The Swiss Civil Code and Code of Civil Procedure, Italian Penal Code, German Code of Criminal Procedure and Code of Commerce, and French Administrative Law were adopted with due adaptations some ninety years ago by the new Turkish Republic (Orucu, 2000). Substantial amendments and modifications have been effected since, particularly in the last decade within the framework of the EU accession process.

5.2.3 Energy and environment

Turkey is a country of 814,000 km^2 having a population of over 70 million. Turkey's territory lies where the three continents making up the old world (Europe, Asia, and Africa) are closest to each other, and where Europe and Asia are separated by the Bosphorus. Turkey's neighbors to the west are Greece and Bulgaria, and to the east Georgia, Armenia, Syria, and the energy-rich Iraq and Iran. In spite of its possession of some limited energy resources, Turkey has to import a greater part of the energy it consumes. Meeting the energy demand by domestic production decreased from 48.1 percent in 1990 to 26.9 percent in 2006 (Ministry of Energy and Natural

Resources, 2009). Turkey has to continue its ever-increasing energy import to keep up with its rapidly growing energy deficit. In this sense, the steps Turkey will take to foster energy efficiency are paramount in their prospective contribution to the national economy. Figure 5.1 shows the distribution of energy consumption among various sectors.

Within the total energy consumption, industry has a share of 40 percent, domestic consumption (in households) 31 percent, transportation 19 percent, and agriculture 5 percent (Ministry of Energy and Natural Resources, 2009). The relatively high share of domestic consumption points to its relative importance as an issue in the management of the energy deficit. Decreasing the share of domestic consumption through efficient use of energy according to green building principles is likely to be an effective remedy in decreasing Turkey's total energy consumption and therefore its dependence on energy imports. Distribution of primary energy consumption among various sources is given in Figure 5.2 (Ministry of Energy and Natural Resources, 2009).

The relatively large share of fossil fuels in meeting Turkey's energy demand poses a problem with respect to environmental sustainability, not only because the fossil fuels pollute the environment, but also because they cannot be replenished. A large part of CO_2 emissions in Turkey originates from domestic use of fossil fuels. This makes buildings one of the highest contributors to the greenhouse gas emissions that cause climate change, as well as one of the biggest opportunities to reduce these emissions (Erten, 2009).

All in all, the energy-related figures show clearly that even incremental improvements in domestic energy consumption would yield considerable benefits as to sustainability in both economic and environmental capacities. Although transition from coal to natural gas as fuel for the heating of houses in large towns has been a favorable move from an environmental standpoint, boosting efficiency in the use of energy via green building principles has already become an urgent necessity under the current circumstances.

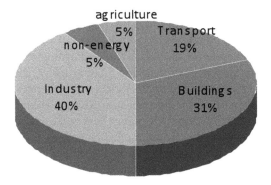

Figure 5.1 Distribution of energy consumption among various sectors in Turkey
Source: Adapted from the Ministry of Energy and Natural Resources, 2009.

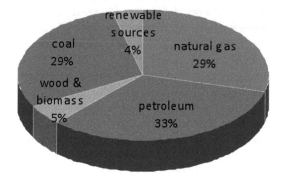

Figure 5.2 Distribution of primary energy consumption among various sources
Source: Adapted from the Ministry of Energy and Natural Resources, 2009.

5.3 Analysis of the current legislation

In addition to the internal factors of energy and environment issues, the external factors of ratifying the Kyoto Protocol and the European Union *acquis*, which Turkey as an accession country is in the process of adopting, have all served to thrust Turkey forward in developing the legislation on sustainability. This is why the analysis of existing regulations commences in this section with a review of the related international treaties.

5.3.1 International treaties

(A) Adoption of the EU acquis

Turkey was officially recognized as a candidate state on an equal footing with the other candidate states at the Helsinki European Council held on 10–11 December 1999. At the end of 2004, the accession negotiations between the EU and Turkey were formally started, initiating a formalized process whereby Turkey is required to complete the adoption of the EU *acquis*. Turkey administers its adoption of the EU *acquis* according to a "National Program" which furnishes a road-map for adaptation of the legislation. The last National Program prepared in 2008 (Secretariat General for EU Affairs, 2009) includes issues regarding the purpose of energy efficiency in buildings. As stated in the National Program, the Regulation on the Energy Performance of Buildings, predicated on the EC Directive on the Energy Performance of Buildings,[2] was promulgated in December 2008. Formulation of the Building Energy Performance Calculation Methodology and development of its software were completed in December 2009. According to the National Program, the following actions will be completed in 2010 in order to support the implementation of this regulation:

- Strengthening of the administrative capacity for implementation of the Regulation.
- Establishment of a laboratory in order to carry out the activities such as R&D, inspection and training regarding the energy performance of buildings.
- Establishment of a "Building Inventory" database for effective implementation of the Regulation.
- Design training courses for the building constructors on ecologic construction, solar energy, zero CO_2 emission buildings and construction techniques for energy efficient buildings.

(B) Kyoto Protocol

Turkey became a party to the United Nations Framework Convention on Climate Change (UNFCCC) in 2004 but refrained from signing the Kyoto Protocol for a prolonged period. On February 5, 2009 a law of conformity numbered 5836 was enacted and the protocol was signed on May 7, 2009. With this Protocol, Turkey has accepted to promote sustainable development,[3] enhance energy efficiency in relevant sectors of the national economy,[4] research on and promote development and increased use of new and renewable forms of energy, of carbon dioxide sequestration technologies and of advanced and innovative environmentally sound technologies[5], encourage appropriate reforms in relevant sectors aimed at promoting policies and measures which limit or reduce emissions of greenhouse gases,[6] and ensure that emissions of the greenhouse gases do not exceed the assigned amounts, calculated pursuant to the quantified emission limitation and reduction commitments.[7] Undoubtedly, promoting greener buildings is a very effective and indispensable way of achieving these commitments. Below is the analysis of the current national legislation regarding common green building themes, most of which have been promulgated in the last five years.

5.3.2 National legislation

Turkey is one of the few countries that included environmental protection in its Constitution. According to the Constitution,[8] everyone has the right to live in a healthy, balanced environment. It is the duty of the state and citizens to improve the natural environment and to prevent environmental pollution. Despite this provision, no specific legislation exists in Turkey currently on green buildings; therefore, a conceptual framework has been developed for reviewing the current legal situation. In this framework, the performance criteria contained in the green building evaluation systems used worldwide (Leadership in Energy and Environmental Design (LEED), Building Research Establishment Environmental Assessment Method (BREEAM) and Sustainable Building Tool (SBTool)) have been categorized and common themes relevant to green buildings were determined. Laws and regulations falling

Table 5.1 Common themes for analysing legislation regarding the green buildings

LEED	BREEAM – The code for sustainable homes	SB tool	Common themes
energy and atmosphere	energy efficiency/ CO_2	energy consumption environmental impact	(1) ENERGY & CO_2
water efficiency	water efficiency surface water management	resource consumption	(2) WATER
materials and resources	use of materials		(3) MATERIALS
sustainable sites		site selection	(4) SITE
	site waste management household waste management		(5) WASTE
indoor environmental quality		indoor air quality	INDOOR AIR QUALITY[1]
innovation and design process		social economical and cultural factors	OTHER

Note
1 "Indoor air quality" is analysed within the "Energy" theme.

under the scope of each theme are dealt with below, also with an eye to identifying the need for new legislation.

The building industry is responsible for a large part of the world's environmental degradation as buildings converge in themselves major indexes of energy and water consumption, raw material employment, and usage of land (Melchert, 2007). This is why use of natural resources like energy, water, land, and raw materials is among important criteria employed in assessment systems for green buildings. On the other hand, buildings generate considerable quantities of waste during their construction, utilization, and demolition, a fact that makes waste management control another crucial theme associated with green buildings.

(A) Energy

Energy efficiency is defined as minimization of energy consumption without compromising either the quantity or the quality of production, nor entailing any impairment in economic growth or degradation in social welfare (Olgun *et al.*, 2009). Legal regulation has recently been made in Turkey with the purpose of stimulating energy conservation, rendering some practices compulsory and enabling inspection of obtained results. These legal measures are as follows:

(A) ENERGY EFFICIENCY LAW

Law 5627 of 2 May 2007 targets increasing the energy resources and the efficiency in the utilization of energy with the purpose of protecting the environment, the effective utilization of energy, cutting energy wastage, and relieving the burden of energy costs on the economy.[9] Industrial plants, electrical energy production facilities, transmission and distribution networks, and transportation are within the purview of this Law, in addition to households.[10] The Law describes the scope of tasks and authority for the institutions and agencies involved in the conservation of energy[11] and regulates the education and consciousness-raising activities for augmenting the effectiveness of energy efficiency services and the energy consciousness.[12] In order that measures are devised to improve energy efficiency in industrial plants, service buildings, public buildings, and commercial buildings with total construction areas of 20,000 m² or more, this Law has made employment or outsourcing of certified energy managers mandatory.[13] Whereas residential buildings are not under an obligation to employ energy managers, the principles of energy performance requirements for such buildings are stipulated to conform to "The Regulation for Energy Performance of Buildings." Within the scope of the Law, subsidies to be granted to research and development and implementation projects have been formulated,[14] as well as cases requiring imposition of administrative sanctions.[15] Secondary legislation based on this Law, aiming at the improvement of building energy performance is as follows.

(B) REGULATION ON EFFICIENT USE OF ENERGY AND ENERGY SOURCES

Regulation 27035 of October 25, 2008 covers various details related to the implementation of issues like energy management, energy inspection, efficiency improvement projects, education and certification, authorization of agencies and companies as associated with these issues, and incentives for efficiency improvement projects.

(C) REGULATION ON ENERGY PERFORMANCE OF BUILDINGS (BEP-Y)

Regulation 27075 of December 5, 2008, also based on the Energy Efficiency Law no. 5627 of May 2, 2007, covers the computational rules for the assessment of all energy uses in a building, and its classification with respect to primary energy use and CO_2 emission, the specification of minimum energy performance requirements for new or substantially renovated buildings, guidelines for the evaluation of the practicability of renewable energy sources, inspection of heating and cooling systems, abatement of greenhouse gas emissions, and the performance criteria and the implementation guidelines that apply to the buildings.[16] This regulation is promulgated on the EC Directive on the Energy Performance of Buildings, which lays down requirements in regard to the general framework for a methodology of calculation of the integrated energy performance of buildings, the application of minimum requirements on the energy performance of new buildings and large existing buildings that are subject to major renovation, as well as the energy certification and regular inspection of buildings and its systems.[17] Below is an overview of the requirements of this regulation.

Scope:
- The regulation applies to the design of new buildings, and to repair and restoration work or electrical and mechanical refurbishment carried out on existing buildings, substantial in scope so as to necessitate a design revision.[18]
- Operation or production buildings in industrial areas, buildings with planned service life less than two years, buildings with total utilization area less than 50 m^2, and buildings with no heating or cooling requirements are outside the scope of this regulation.[19]

Architectural principles:
- In the architectural design, maximum use should be made of possibilities of natural heating, cooling, ventilation and lighting in order to minimize the heating, cooling, aeration and lighting requirements, giving due consideration to sun, humidity and wind factors.[20]
- In the orientation of the building and of the internal spaces, the local meteorological data should be taken into consideration in order to minimize unintentional heat gain or loss,[21] and the main living spaces within the building should be located so as to make optimum use of sunlight, solar heat and natural draft.[22]
- The architectural details should include junctions of the construction elements so as to show continuity of thermal insulation.[23]
- The reports on the possibility of making use of renewable sources of energy should be considered before all else in the architectural design process.[24]

Thermal insulation principles:
- The annual heating energy requirement of the building should be less than the limit value indicated in TS 825 (Turkish Standards Institute, 2009).[25]
- The outer shell of the building should be insulated in compliance with TS 825 (Turkish Standards Institute, 2009) avoiding formation of heat bridges.[26]

- The "thermal insulation project" should be prepared by the authorized mechanical engineer and submitted to the administration during application for the construction licence.[27]

Mechanical principles:
- The mechanical installation plumbing and equipment to be used in the heating, cooling, ventilation and air conditioning systems should be insulated with heat and/or sound insulation materials.[28]

Indoor air quality principles:
- Appropriate materials should be used to provide impermeable air tightness in surfaces, sections and/or shafts where transfer of heat is likely. Provision should be made to maintain the required amount of air recirculation to prevent deterioration of indoors air quality.[29]

Heating principles:
- A central heating system should be installed in a new building with a total utilizable area exceeding 2,000 m^2.[30]
- In buildings with central heating systems, central or local heat or temperature control equipment and systems should be used to allow distribution of heating costs according to the quantity of heat usage.[31]

Cooling principles:
- Central air conditioning systems should be designed for non-residential buildings with a total cooling power requirement of 250 kW or greater.[32]

Ventilation and Air-Conditioning Principles:
- Air conditioning systems should have appropriate mechanical equipment to enable operation at varying air flows, maintaining control of indoors air at varying human loads.[33]

Hot water preparation principles:
- Designing central sanitary hot water preparation systems is obligatory for hotels, hospitals, student hostels, and similar boarding, non-residential buildings and sports centres with utilizable areas over 2,000 m^2.[34]
- The design maximum water temperature in central sanitary hot water systems should not exceed 60°C.[35]
- The equipment and distribution plumbing in central sanitary hot water systems should be thermally insulated. However, water temperature should be kept at 60°C for an hour every week in order to prevent legionella.[36]

Electrical installation and lighting principles:
- According to the type of building and purpose of space utilization, systems and switches (timer-controlled relays, etc) should be used as appropriate

to enable maximum utilization of sunlight and avoid unnecessary artificial lighting.[37]

- To allow efficient use of electrical energy, high-efficiency lighting equipment should be used, controlled by light sensors where appropriate, and light colours be preferred on indoor walls and the furniture.[38]

Automation principles:
- Buildings with central heating, air conditioning and/or cooling systems should be equipped with control and automation equipment that allows separate adjustment of each room temperature.[39]
- Control of lighting in buildings is done according to time, daylight, and usage, except in residential buildings.[40]
- Installation of a computer controlled building automation system is mandatory for buildings with utilization areas greater than 10,000 m^2 and where central heating, air conditioning, ventilation, and lighting systems are all present.[41]
- In new buildings, an energy monitoring system should be installed that allows separate measurement and registering/reporting of the energy consumption by the heating, air conditioning, and ventilation systems.[42]

Renewable energy and cogeneration principles:
- In new buildings with utilizable areas over 20,000 m^2, the use of renewable energy air, water, or soil source heat pump; co-generation and micro-generation systems should be analyzed by the designers to partially or totally meet the energy requirements of the building. One or more of these systems should be realized in the project and the cost of these systems should exceed 10 percent of the project cost as calculated with the unit prices published by the Ministry of Public Works and Settlement.[43]

Energy Performance Certificate:
- Energy Performance Certificate is issued in accordance with the Building Energy Performance Calculation Methodology (BEP-HY).[44]
- Energy Performance Certificate is valid for 10 years following date of issue.[45]
- Energy Performance Certificate is issued by the authorized agency and approbated by the administration. This certificate is an inseparable part of the building utilization licence for the new buildings.[46]
- A copy of the Energy Performance Certificate is retained by the owner, superintendent, managing board, and/or energy manager of the building and another copy is displayed in a visible place in the entrance of the building.[47]
- A new Energy Performance Certificate is issued within one year, in compliance with this regulation, in case of modifications that alter the annual energy requirement of the building.[48]

- Energy Performance Certificate should be issued for the entire building. Upon request, certificates may be issued separately for each individually owned dwelling unit or various differently utilized spaces.[49]
- Energy Performance Certificate is issued by using the Building Energy Performance Software Tool (BEP-TR). Only certified energy experts and companies have access to this software.[50]
- The Ministry of the Public Works and Settlement publishes the criteria which will apply to the training and examination of the energy experts.[51]
- When the building or one of its parts is sold or let, the Energy Performance Certificate is handed over by the landlord to the tenant or new owner.[52]
- In the Energy Performance Certificate are indicated the total energy consumption and the separate energy consumptions for the heating, sanitary hot water production, air conditioning, and lighting systems (primary and ultimate), beside the energy consumption and greenhouse gas emission classes of the building as well as information regarding the use of renewable energy in the building.[53]
- Energy Performance Certificates are issued within ten years following coming into force of the Energy Efficiency Law, for the existing buildings or buildings under construction not yet issued a building utilization licence.[54]

Periodical controls and reporting:
- All systems prescribed in the Regulation are subjected, under the responsibility of the owner, superintendent, managing board, and/or energy manager of the building, to periodic control, testing, and maintenance in accordance with the reports prepared in the design stage and approved by authorities so as to keep the building's energy performance as designed and approved.[55]

Sanctions:
- A construction licence is not granted if the architectural, mechanical and electrical designs of the building do not comply with the specifications as set forth in the Regulation.[56]
- In case the designs compliant with the Regulation are not fully implemented, the building is not granted a building utilization licence by the relevant administration, until the identified defects are made good.[57]

It has been declared that the method of computing the energy performance as prescribed in the Regulation on Energy Performance of Buildings will be announced until 1 January 2011.[58] An introduction of the mentioned computational method and the associated software was made by the Ministry of Public Works and Settlement at a meeting held on December 19, 2009. Below are brief explanations of the calculation methodology and the software.

(D) BUILDING ENERGY PERFORMANCE CALCULATION METHODOLOGY (BEP-HY)

Regulation on Energy Performance of Buildings requires the evaluation of the energy performance of buildings including determination of energy consumption, CO_2 emission, benchmarking these values to a reference building and classification of the building according to the benchmark results. Building Energy Performance Calculation Methodology (BEP-HY) defines the assumptions and methods used in the calculation of energy consumption of heating, cooling, ventilation, lighting, and hot-water systems in a building. BEP-HY uses hourly climatic data and is based on a simple hourly dynamic method, which calculates the net energy demand and energy consumption of the building systems on an hourly basis. EN standards such as the EN ISO 13790, Turkish standards, and ASHRAE are used in BEP-HY.

In BEP-HY, buildings are classified as residential, office, educational, healthcare, hotel, and shopping with regard to their typology and as rectangular, cross, courtyard, etc. with regard to their geometry. BEB-HY uses zoning in the calculations; every unit in an apartment block and the circulation area are regarded as different zones in residential buildings. In complex buildings, such as healthcare or educational buildings, every floor is taken as a different zone. BEP-HY can be used for comparing the energy performance of alternatives in the design phase, comparing the cases of using and not using energy efficiency measures, developing foresights for the need of energy in future by calculating typical buildings that can represent the building stock in an area, and developing a national database of energy performance of various building elements and materials (Ministry of Public Works and Settlement, 2009).

(E) BUILDING ENERGY PERFORMANCE SOFTWARE TOOL (BEP-TR)

BEP-TR is an internet-based application of the national calculation methodology that produces the building energy certificate based on the data entered by the expert. The software was developed with the aim of centralizing the collected data on the energy performance of buildings and monitoring the energy certification process. The use of the software includes the steps below (Ministry of Public Works and Settlement, 2009):

1   Certified energy performance experts are given a user name and password to be able to access and use this software.
2   The expert enters the data about the building and her/his calculations.
3   These data are transferred to the central database through the internet.
4   Necessary calculations are done by the software according to the calculation methodology (BEP-HY).
5   Results are benchmarked with a reference building by the software. (Reference building is assumed to have the same settling features, same geometry – plan, roof type, number of floors and total area, mechanical,

hot water and lighting systems that meet the minimum requirements of the current legislation, no renewable energy or cogeneration systems, and a building shell that is compatible with the TS 825 standards).

6   Energy and CO_2 emission class of the building is determined according to the benchmark results and the energy certificate for the building is prepared by the software and sent to the expert by e-mail.

After the installation of BEP-TR to the central server and development of its interface, the tutors and the certification experts will be trained and energy certification will begin (Bayram, 2009).

Criticism about the statutes on energy efficiency mainly focuses on the issues of financing, incentives, awareness building, and education standards in the certification process. It is prescribed in the existing legislation that the certificate courses for the experts, who will be responsible for issuing the building energy certificates, will be held by the universities and chambers of engineers, and also by the companies they will authorize.[59] Olgun *et al.* (2009) argue that considering the probable deficiencies as to expert personnel and laboratory facilities of such prospective companies, they may fail to achieve equivalent quality and standards in education. Olgun *et al.* (2009) also suggest that additional legislation is needed for financial support of efficiency improvement projects and energy efficiency control studies. Although existing legislation prescribes partial state financing for energy efficiency projects in industrial plants,[60] incentives for energy efficiency projects in residential buildings and consciousness-building among the financial corporations may be a decisive step in the provision of necessary financial resources for relevant progress.

In his evaluation of the Building Energy Performance Regulation Camlibel (2009), the CEO of a prominent Turkish construction company, draws attention to the importance of incentive mechanisms in effective implementation of such regulations. Camlibel (2009) suggests high initial investment costs associated with the energy efficiency systems will have negative effects on implementation of the statutes and reminds that due to cost concerns only 15 percent of new buildings built after promulgation of the Regulation of Thermal Insulation in Buildings comply with the requirements of this regulation. Mechanisms, such as tax reduction on the equipment related to energy efficiency, reduction on VAT, construction licence duty and estate duties for the projects in which renewable energy and cogeneration systems are used, should be in place. According to Camlibel (2009) another important factor in the successful implementation is awareness-building activities. He suggests that it is possible to make a 15 percent saving on energy just by building the energy efficiency awareness in the public. Professional associations (such as the Turkish Green Building Council), NGOs, public and private institutions have to work together to build awareness of energy efficiency among building owners, users, designers, construction companies, and material suppliers.

(B) Water

One of the most serious problems that Turkey may face in parallel with global warming is water shortage. Turkey's annual water capacity is 1,430 m³ per capita and this quantity is well below the 7,600 m³ per capita world average (Soyak, 2009). Turkey has lost half of its wetlands in the last forty years and has to urgently take measures to preserve existing ones. Undoubtedly, efficiency in residential water consumption is one of the most effective measures for this purpose, yet currently no legislation exists in Turkey regarding water efficiency in buildings. Such legislation could address the following issues:

- Use of efficient water equipment in buildings, such as faucets with photocell devices, air-mixed shower heads, flush mechanisms that provide different amounts of water, provide control to stop the water flow and consume less than 6 liters of water in each use and shower cabinets instead of bath tubs.
- Use of rainwater harvesting systems in buildings with a roof area exceeding a certain limit.
- Use of surface water harvesting systems in suitable projects.
- Where the building has an independent wastewater treatment plant, recycling of the treated effluent for appropriate reuse, like lawn irrigation.
- Use of efficient irrigation systems.
- Use of plant types that consume less water in landscaping.
- Water efficiency awareness-building programs for the public.

(C) Building materials

Building materials are assessed in terms of recyclability and the level of toxic contents in green building certification systems. Regulation on Building Materials requires the materials used in buildings to be mechanically resistant, non-toxic, safe to use, energy efficient[61] and have a CE or G marking.[62] This regulation could be amended to include limits and/or incentives regarding the environmental impact of the production process of materials using the following criteria:

- The amount of energy/water consumed in the production of material
- The amount of harmful emissions created during production
- Ratio of renewable energy used in production
- Ratio of recyclable material used in production
- Ratio of the local raw material used in production (to decrease energy use and emissions related to transportation)
- Ratio of the renewable raw materials used in production

Furthermore, additional legislation could be prepared to encourage the use of "green materials" in buildings such as recycled, recyclable, local, and renewable material. Such legislation might be supported with training of designers and professionals dealing with material selection in construction projects.

(D) Site

Development plans are the main source of restrictive regulations in site selection for buildings. Planning areas are divided into zones such as residential, commercial, industrial, etc. and some development restrictions such as the maximum number of floors and the usable area to land area ratio are included in these development plans. Ercoskun (2005) argues that current development plans do not consider urban identity and ecological values such as climate, spaces between buildings, direction, natural lighting, and air circulation, etc., resulting in the Turkish cities to be unsustainable. Furthermore, the reality of illegal settlements and brutal development interests has a damaging effect on both natural and built environment in big cities.

However, limited to large-scale projects, a key piece of legislation regarding environmental concerns in site selection is the Regulation on Environmental Impact Assessment, requiring an independent assessment report evaluating the positive and negative impacts of a project on the environment and appropriateness of the site selected for the project.[63] The scope of this regulation covers energy production and distribution plants, industrial and agricultural buildings, as well as hotels with 100 rooms or more, public and private housing projects with 200 units or more, educational and sports complexes.[64]

(E) Waste

Buildings produce waste in all phases of their life cycle. Regulation on the Control of Excavation, Construction and Demolition Wastes, promulgated on 13 March 2004, sets the rules for the collection, accumulation, recycling, and disposal of wastes generated in the excavation, construction, and demolition phases.[65] According to the regulation, wastes should be minimized at the source, the persons, or organizations who are responsible for the management of the wastes should take the necessary precautions to minimize the harmful impacts of the wastes to the environment, the wastes should be recycled and used as building materials where possible, excavation soil and construction/demolition wastes should be processed separately, wastes should be classified at the source for a healthy recycling process, the producers of the unrecyclable excavation, construction, and demolition wastes should pay for the disposal of such wastes.[66] Regarding the recycling of excavation soil, the regulation requires the vegetative soil to be used in parks etc. and the rest to be used in filling, as a cover material in solid waste storage areas, and as clay in cement industry where the chemical properties of the soil allow.[67] Collection, treatment, disposal, and recycling of wastewaters and solid wastes produced during the utilization of buildings, i.e. wastewater and solid waste management are also well-established fields covered by separate special legislation and institutional agencies.

An overall evaluation of the current legislative and institutional framework regarding sustainability has also been made in the Ninth Development Plan of the Prime Ministry covering the years 2007 to 2013, which suggests that

> within the EU harmonization process, even though progress has been achieved regarding waste management, protection of the nature, noise and environmental impact assessment, many regulations are still needed. The uncertainties about the duty and authority distribution among institutions, for the sustainable use of natural resources while protecting the environment without adversely affecting the production process, could not be adequately eliminated. There is still need to improve the institutional framework for environmental monitoring, auditing and reporting for increasing efficiency in these tasks and to ensure that the information flow and exchange among the relevant institutions are carried out through an integrated system. Capacity of municipalities about the planning, designing, implementation and operation of environmental infrastructure services need to be improved. In the framework of the conditions of Turkey, and with the participation of the relevant parties, a National Action Plan that sets forth the policies and measures for reducing greenhouse gas emissions will be prepared. Thus, responsibilities concerning UN Framework Convention on Climate Change will be fulfilled.
>
> (State Planning Organisation, 2006).

5.4 Conclusion

In this chapter, the current legislation in Turkey has been analyzed in the framework of "green building themes" (energy, water, materials, site, and waste) developed on the basis of performance criteria used in the green building certification systems. The following deficiencies in policies and statutory regulation have been pointed out regarding the legislation, and proposals put forward for the promotion of green buildings in Turkey.

Energy efficiency is the most advanced theme and there is well-established legislation in this area, including the Energy Efficiency Law, Regulation on Efficient Use of Energy and Energy Sources, Regulation on Energy Performance of Buildings (BEP-Y), Building Energy Performance Calculation Methodology (BEP-HY) and Building Energy Performance Software Tool (BEP-TR). BEP-Y sets the architectural, thermal insulation, mechanical, indoor air quality, heating, cooling, ventilation, air-conditioning, hot water preparation, electrical installation, lighting, building automation, renewable energy, co-generation, and certification principles for energy efficiency in buildings. Criticism about the statutes on energy efficiency mainly focuses on the issues of financing, incentives, awareness building, and education standards in the certification process.

Currently no legislation exists in Turkey regarding water efficiency although one of the most serious problems that Turkey may face in parallel with global warming is water shortage. It is suggested that a water efficiency regulation should be prepared and include provisions regarding the use of efficient water equipment, rainwater harvesting systems, surface water harvesting systems, efficient irrigation systems, plant types that consume less water in landscaping, recycling of the treated effluent for appropriate reuse, and water efficiency awareness-building programs.

Building materials are among important criteria employed in assessment systems for green buildings due to major indexes of raw material consumed in buildings, as well as the natural resources consumed during the production of building materials. With the aim of limiting this impact, it is suggested that the Regulation on Building Materials should be amended to include limits and/or incentives regarding the environmental impact of the building materials' production process. A list of criteria is also suggested for determining this impact such as the amount of energy/water consumed, harmful emissions created, renewable energy used, and recyclable material and local raw material used in the production of material. Furthermore, it is also suggested that additional legislation should be prepared to encourage the use of "green materials" in buildings such as recycled, recyclable, local, and renewable material by designers and professionals dealing with material selection in construction projects.

Site selection is the most complex theme requiring regulations and resolute policies on various scales. Besides restoring and implementing development plans with an ecological concern, the problem of illegal settlements in big cities needs to be solved in order to set forth a realistic approach to the issue of sustainable buildings and cities in Turkey. Sites that have current access to public transportation and infrastructure should be preferred for new developments.

Buildings produce waste in all phases of their life cycle. Regulation on the Control of Excavation, Construction and Demolition Wastes sets the rules for the collection, accumulation, recycling, and disposal of wastes generated in the excavation, construction and demolition phases. Collection, treatment, disposal, and recycling of wastewaters and solid wastes produced during the utilization of buildings are also well-established fields covered by separate special legislation and institutional agencies.

Besides improvement of the current national legislation, appropriate organization of institutions authorized to implement this legislation, as mentioned in the Ninth Development Plan of the Prime Ministry, and resolute policies to support implementation, such as awareness-building and incentives, are also indispensable steps of achieving a "green" built environment in Turkey.

Notes

1 Based on purchasing-power-parity valuation of country GDP.
2 Council Directive (EC) 2002/91 on the Energy Performance of Buildings [2002] OJ L 1/65.
3 Kyoto Protocol Art. 2.
4 Art. 2.1.a.i.
5 Art. 2.1.a.iv.
6 Art. 2.1.a.vi.
7 Art. 3.1.
8 The Constitution 1982 s 56.
9 Energy Efficiency Law 2007 s 1(1).
10 s 2(1).
11 s 4 and 5.
12 s 6.
13 s 7(1a).
14 s 8.
15 s 10.
16 Regulation on Energy Performance of Buildings 2008 s 1(1) and 2(1).
17 Council Directive (EC) 2002/91 on the Energy Performance of Buildings [2002] OJ L 1/65, Article 1.
18 Regulation on Energy Performance of Buildings 2008 s 5(1).
19 s 2(2).
20 s 7(1).
21 s 7(2a).
22 s 7(2b).
23 s 7(2c).
24 s 7(2ç).
25 s 9(1a).
26 s 9(3).
27 s 10(1).
28 s 11(1).
29 s 12(1).
30 s 13(3).
31 s 13(7).
32 s 15(1).
33 s 17(5).
34 s 19(3).
35 s 19(5).
36 s 19(9).
37 s 21(2).
38 s 21(8).
39 s 20(2).
40 s 20(5).
41 s 20(6).
42 s 20(8).
43 s 22(1).
44 s 25(1).
45 s 25(2).
46 s 25(4).
47 s 25(6).
48 s 25(7).
49 s 25(8).
50 s 25(13).

51 s 26A(1).
52 s 25(15).
53 s 26(1).
54 s Temp.3.
55 s 24(2).
56 s 5(2).
57 s 5(3).
58 s Temp.2.
59 s 26A(1).
60 Regulation on Efficient Use of Energy and Energy Sources 2008s 15–17.
61 Regulation on Building Materials 2002 App. 1.
62 s 6.
63 Regulation on Environmental Impact Assessment 2008 s 4(1c).
64 Regulation on Environmental Impact Assessment 2008 App. I and II.
65 Regulation on the Control of Excavation, Construction and Demolition Wastes 2004 s 1.
66 s 5.
67 s 26.

References

Akbiyikli, R., Dikmen, S. U. and Eaton, D. (2009) "Sustainability and the Turkish Construction Cluster: A General Overview", RICS COBRA Research Conference, 10–11 September, Cape Town, South Africa.

Bayram, M. (2009) "Bina Enerji Performansı Süreçleri" available at http://www.yigm.gov.tr/bayindirlik/enerji_kimlik_belgesi.pdf (accessed January 2010).

Camlibel (2009) "Emre Çamlıbel: Yönetmeliği Desteklememek Mümkün Değil, ancak ..." available at http://www.yapi.com.tr/Sektorden/emre-camlibel-yonetmeligi-desteklememek-mumkun-degil-ancak_74794.html (accessed January 2010).

Engineering News-Record (2009) "ENR Top 225 International Contractors" available at http://enr.construction.com/toplists/InternationalContractors/001-100.asp (accessed December 2009).

Ercoskun, O.Y. (2005) "Sustainable City Plans against Development Plans", *Journal of Science* (Gazi University), Vol. 18(3), 529–544.

Erten, D. (2009) "Financial Incentives to Accelerate the Construction of Green Buildings in Turkey and Global Best Practices", available at www.uevf.com.tr/uev f1/sunumlar/pp06-03.doc (accessed December 2009).

International Monetary Fund (2009) "World Economic Outlook Database" available at http://www.imf.org/external/pubs/ft/weo/2008/01/weodata/weoselgr. aspx (accessed December 2009).

Katsarakis, Y., Rezk, A., Sazak, E., Shaydullin, H., and Yadikar, B., (2007) "Turkey and the construction services cluster", working paper, Institute for Strategy and Competitiveness, Harvard Business School, Boston, Spring Term.

Melchert, L. (2007) "The Dutch Sustainable Building Policy: A Model for Developing Countries?", *Building and Environment*, Vol. 42, 893–901.

Ministry of Energy and Natural Resources (2009) "Statistics", available at http://www.enerji.gov.tr/index.php?dil=tr&sf=webpages&b=y_istatistik&bn=244&hn=244&id=398 (accessed November 2009).

Ministry of Public Works and Settlement (2009) "Binalarda Enerji Performansı", available at http://www.yigm.gov.tr/bep/sunumlar/bep.pdf (accessed January 2010).

Olgun, B., Kurtulu_, O., Gültek, S., and Heperkan, H. (2009) Enerji Verimliliği ve Türkiye'deki Mevzuat, IX. Ulusal Tesisat Mühendisliği Kongresi (TESKON 2009), _zmir, Türkiye.

Orucu, E. (2000) "Critical Comparative Law: Considering Paradoxes for Legal Systems in Transition", *Electronic Journal of Comparative Law*, Vol. 4.1, available at http://www.ejcl.org/41/art41-1.html (accessed November 2009).

Secretariat General for EU Affairs (2009) "National Programme of Turkey for the Adoption of the EU Acquis", available at http://www.abgs.gov.tr/index.php?p=42260&l=2 (accessed December 2009).

Soyak (2009) "Sustainable Living", Mikado, Istanbul.

State Planning Organisation (2006), Ninth Development Plan: 2007–2013, State Planning Organization, T.R. Prime Ministry, Ankara-Turkey, available at http://ekutup.dpt.gov.tr/plan/ix/9developmentplan.pdf (accessed December 2009).

Turkish Contractors Association (2009) available at http://www.tmb.org.tr/index.php?l=eng (accessed December 2009).

Turkish Green Building Association (2009) available at http://www.cedbik.org (accessed December 2009).

Turkish Standards Institute (2009) "TS 825", available at http://www.tse.org.tr/TSE IntWeb/Standard/Standard/StandardAra.aspx (accessed December 2009).

Turkish Statistical Institute (2009) available at http://www.turkstat.gov.tr/Pre IstatistikTablo.do?istab_id=1002 (accessed December 2009).

US Green Building Council (2009) "Green Building Research", available at http://www.usgbc.org/DisplayPage.aspx?CMSPageID=1718 (accessed December 2009).

WCED (1987) "Our Common Future", available at http://www.un-documents.net/wced-ocf.htm (accessed December 2009).

World Trade Organization (1998) available at http://www.wto.org/english/tratop_e/tpr_e/tp83_e.htm (accessed December 2009).

6 Surveying the sustainable and environmental legal and market challenges for real estate

Colleen Theron and Malcolm Dowden

6.1 Introduction

It has been estimated that the built environment in its widest sense (including construction) is responsible overall for about 40 per cent of carbon dioxide emissions, as well as 40 per cent of all energy used. Commercial property is a major contributor to this.

Despite the recent negative publicity challenging the scientific evidence underpinning climate change, the consensus seems to remain that climate change must be tackled and cutting carbon emissions is critical to achieving this. As a result (and notwithstanding the global economic downturn and the depressed property market), legislation and policy developments at global and national level has continued to crystallise solutions for reducing greenhouse gas emissions (GHG), including the built environment. In the United States, states and local governments have embraced green real estate initiatives, ahead of federal legislation. Since 2005 there has been an increase in approved green building polices from 13 to 31.

In the UK, the drive towards a low carbon economy has led to a number of policy and legislative initiatives to tackle the reduction of carbon in buildings, such as the low carbon transition plan and the consultation on zero carbon for non domestic buildings. The agenda has also been driven by directives from the European Union, which member states are bound to implement. Owners and managers of existing commercial property buildings are finding themselves bound by a raft of new environmental legislation, such as energy efficiency standards for buildings; a mandatory carbon emission reduction scheme (where the organisation produces electricity above a threshold of 6000kM per half hour); and revised building regulations.[1] Legislative and market pressure is also being put on the property industry to design, construct, use and demolish buildings in a way that both mitigates and adapts to climate change. As the impact of climate change and resource pressures deepen, the effect on asset value will increase. The creation of the coalition government in the UK has not deterred from the theme of a 'green economy', including incentivisation for 'green growth' and 'decarbonisation' of the economy.[2]

This chapter examines how environmental and sustainability issues are challenging 'business as usual' models in the real estate sector, outlining key mandatory and voluntary requirements for 'green buildings'.

6.2 Is sustainability green?

There is widespread and considerable interest in the topic of sustainability and sustainable construction. However, there is also a good deal of confusion over the terms used. The term 'sustainable development' is often confused with environmental protection, and many limit the scope of sustainability to consideration only of environment issues (Schleich *et al.*, 2009[3]).

The concept of 'sustainable development' is not always interpreted in the same way by different countries. The standard definition is the one contained in the Brundtland Report, namely

> *Sustainable development that meets the needs of the present without compromising the ability of future generations to meet their needs.*[4]

'Sustainability' as a legal term has its origins in the 'soft law' of international conventions. This means that it is too vague for breach to give rise to legal sanction (Keay, 1990). In the UK the statutory guidance issued to the Environment Agency to supplement the Environment Act 1995 adopts the 'Brundtland' definition of sustainable development.

The concept of 'sustainability' is also often used interchangeably with the concept of Corporate Social Responsibility (CSR) at corporate levels, but clarity on the differences is important for the way business behaves. The European Union started to develop the concept of CSR in 2000/2001 in line with the strategy adopted in Lisbon in 2000. The so called Lisbon objective called on the EU to become the foremost economy in the world, focusing on sustainable economic growth and greater social cohesion by 2010 (Loew *et al.*, 2004). The EU's sustainability strategy is also tied up with the Lisbon objective. From a business perspective CSR came first and was primarily concerned with social matters. Sustainable development emerged from the environmental protection debate. The Brundtland Commission concluded that social, ecological and economic concerns must be given equal weight. CSR tends to be restricted to ecological and social challenges and economic contributions to sustainability are not considered in detail. Whilst the concepts overlap they are applied differently.

Sustainability, in the context of the business world, is used to refer to how environmental, social and economic considerations are integrated into corporate strategy and capital markets for the long term. Companies, historically, have separated financial issues from non-financial aspects of their business. This started changing with the onset of increased pressure by stakeholders and society for more transparency about governance and the impact that companies have on their surrounding environment. Ceres[5] states that the 'license to operate' can no longer be taken for granted by business. Companies, they suggest, cannot consider sustainability challenges in isolation. A recent Pricewaterhouse Coopers survey of 140 chief executives of US-based multilateral companies found that 85 per cent of them believe that sustainable development will be even more important to their business model in five years time than it is today. This should extend to real estate investors and major developers.

In the real estate sector, environmentally sustainable buildings are also referred to as 'green buildings'. Although, 'green buildings' have been used as a synonym for sustainability, this has often led to sustainability being understood from an environmental perspective, neglecting the social and economic perspectives. Despite the lack of a universally agreed definition of 'sustainable buildings', as the market evolves and as new metrics and regulation are developed and implemented, consensus may emerge. The practice of 'green building' is seen as the practice of creating structures and using processes that are environmentally responsible and resource efficient throughout a building's life cycle from siting to design, construction, operation, maintenance, renovation and deconstruction. There is an assumption that sustainable property will perform beyond the baseline of compliance and that it will offer considerable benefits over and above conventional or merely compliant property.

6.3 What are the drivers?

A major driver for sustainability in the real estate sector has been the growth in environmental legislation to regulate the impact a property has on the environment through its whole life cycle. The overall increase in legislation at both international, EU and domestic level does not show signs of any abatement, which will continue to drive change. It is believed that the environmental and social aspects of sustainability will also impact on property performance.

However, there are relatively few studies that have been undertaken about the business case to promote sustainable development in the real estate sector. In 2010 the RICS Foundation, in conjunction with Kingston University, London issued a report entitled "Is sustainability reflected in commercial property process: an analysis of the evidence base'.[6] The report assessed the evidence that exists on whether sustainability is reflected in commercial property prices. It found that although the last ten years has shown an increase in reports on the business case for sustainable property, very few large-scale empirical studies have been undertaken. The greatest number of reports are US based. The conclusions in the report emphasise the need for a clear definition of sustainable buildings and what sustainability features really matter to tenants and building occupiers. The provision of meaningful benchmarks will be able to support valuers in the preparation of valuations. To complement this report, the RICS believes that further research is needed in relation to the financial performance of sustainable buildings.

A joint study undertaken by the University of Regensburg and Helsinki University of Technology (Schleich Report) identifying the drivers of sustainability in real estate practice, confirms the limited availability of research in this area. Whilst some mention is made of the corporate drivers, the study focused on three main property-led drivers, namely increased rental level, decreased property costs and decreased risks.

6.3.1 Increased rental level

Those studies that have analysed the comparative data of energy costs of green buildings to other conventional buildings show reductions of energy use and savings between 6 and 30 per cent.

The Schleich Report highlights that surveys conducted by Jones Lang LaSalle[7] (2008) amongst corporate occupants globally, and a survey undertaken by Cushman and Wakefield (2009) show signs of a willingness to pay for rental premiums for sustainable space, the range of the size of the premium varying from 2–12 per cent.

It is believed that a tenant's willingness to pay premiums for sustainable real estate is a result of increased occupant productivity, potential image benefits and lower running costs.

However, the evidence of a willingness to pay premium rents is far from conclusive. Perceived benefits such as lower operating costs depend on occupier behaviour and where that is achieved in whole or in part by the imposition of 'green lease' obligations, it may be that increased obligations or more rigorous restrictions on use would in fact blunt any argument that the landlord should receive an enhanced rent.[8]

6.3.2 Lower operating costs

Lower operating costs are considered to be an issue promoting sustainable buildings. Empirical evidence indicates that operating costs of certified buildings are lower compared to non-certified buildings. A study in 2008 by Miller *et al.* indicated lower operating costs on a sample of 243 certified and 2,000 non-certified buildings.

6.3.3 Decreased risks

The decreased risk associated with sustainable buildings arises from the perception of lower vacancy rates; however, this is not conclusive.

6.3.4 Reputation

There is a corporate driver for improving reputation, by publishing CSR reports and Carbon Disclosure reports. The drive towards mandatory reporting across the globe will no doubt see an increase in more information being available about property companies. Also, as more companies embrace 'sustainability' in the way they develop their strategies and reporting, this should lead to more transparent information. Theoretically, such positioning by corporations should translate through market mechanisms into increased demand for sustainable property. Although the principles of sustainability may be embedded in the policies of some property owners and occupiers, translating them into property decisions has been difficult.

6.3.5 External factors

Another external driver is the issue of how much weight asset managers of institutional investors are giving to sustainability and climate-related issues. A study by Jones Lang LaSalle in 2008 sought to make the claim that investing in accredited or sustainable stock either does or may yield higher returns and other benefits.

6.4 Are real estate investors interested in energy efficiency?

A 2010 report 'Energy Efficiency in Real Estate Portfolios'[9] published by Ceres, and Mercer,[10] outlines the business case for investing in energy efficiency on the basis that it enhances value in real estate portfolios.

The report aims to provide direct and indirect real estate investors with the background information, academic and industry research, case studies, key steps and best practices for integrating energy efficiency across their portfolios.

The report also states that fiduciaries responsible for these portfolios may assume unnecessary risk and overlook substantial opportunities to enhance returns if they fail to factor energy efficiency into their real estate investment decisions. A further Ceres report on Asset managers' practices (Ceres, 2010) highlights specific best practices that asset managers are purporting to use to incorporate climate risks into their due diligence, corporate governance and portfolio valuation. However, the report states that there are very few asset managers actually doing this.

The issues raised in these reports highlight that climate risks and opportunities are rising on the investment agenda as environmental, social and economic implications of climate change begin to crystallize.

6.5 Legal standards for sustainability

The life cycle of a commercial property is affected by both legal and voluntary standards for sustainability. Pressure is mounting for all parties involved in real estate to do more to achieve sustainability throughout the lifecycle of a building, from design to demolition.

6.6 Building design

6.6.1 Legal standards

The Building Regulations 2000, as amended (BR2000) set out a minimum sustainability standard relating to, for example, energy and water efficiency. The local authority may refuse to issue a completion certificate if the new buildings fail to meet the BR2000 standards.

The impact of the Flood and Water Management Act 2010 will also require developments to be designed to prevent surface water run-off.

The main voluntary tool for assessing the sustainability of new commercial buildings is the Building Research Establishment Environmental Assessment Method (BREEAM). BREEAM awards new commercial buildings a 'rating' measuring their level of sustainability over and above the minimum requirements of the BR2000. It does not, however, address existing buildings.

The Green Rating[11] is a new alternative to BREEAM. It is an assessment of the energy efficiency of a building that looks at both the building materials and the waste generated by the building. It is an international standard intended to allow companies with an international property portfolio to accurately measure the efficiency of their sites.

Other sustainability assessment tools are LEED (Leadership in Energy and Environmental Design) developed in the United States and Canada; and Green Star and NABERS (National Australian Built Environmental Rating System) developed in Australia.

6.7 The planning system

6.7.1 Local authorities

The UK planning system increasingly requires the promotion of sustainable development in planning applications.

The Planning Act 2004 requires local planning authorities to exercise their planning powers with the objective of contributing to the achievement of sustainable development, taking into account government Planning Policy Statements (PPS) (such as PPS 1 and 22) promoting on-site renewable or low-carbon energy generation when drafting development plans. In their capacity as 'reporting authorities' under the Climate Change Act 2008, local planning authorities also have a crucial role to play in climate change mitigation, adaptation and in facilitating transition to a low-carbon economy.

The result is that developers are increasingly required to 'front load' their planning applications with details at the outline planning stage of sustainability considerations. Developers also have to include sustainability standards for new developments as a planning obligation under Section 106 Agreements.

For local planning authorities, though, demonstrating compliance with increasingly broad and complex statutory duties at a time of significant public sector retrenchment is a major challenge. The UK's coalition government has announced a radical shift towards a localized planning system, abolishing regional planning bodies and policies and allowing local planning authorities to produce local plans within a single broad national framework. This move towards 'local democracy', including enhanced rights for local residents to challenge planning decisions, may exacerbate the difficulties faced by unpopular developments, such as waste and energy from waste projects.[12]

6.7.2 Environmental impact assessment

Under the Town and Country Planning (EIA) (England and Wales) Regulations 1999, planning applications for major developments are required to be accompanied by an environmental statement detailing the likely impacts on the environment, and measures to be taken to mitigate those impacts.

6.8 Construction of new buildings

6.8.1 JCT contracts

The standard form of contract[13] used by the construction industry (drafted by the Joint Contracts Tribunal) has recently been amended to include sustainability clauses. The sustainability clauses and guidance do not impose a rigid set of targets upon parties, but instead create a framework in which a contract can include sustainability criteria.

6.8.2 Waste management

Waste is of increasing environmental and economic significance as the increasing regulatory pressure and rising cost of landfill taxes is making waste management a significant cost issue for many organizations.

The Site Waste Management Regulations 2008 require developers of projects worth over £300,000 to prepare a Site Waste Management Plan (SWMP). The SWMP will record details of the construction project, estimate the types and quantities of waste that will be produced and confirm the actual waste types generated and how they have been managed.

Buildings also have to be constructed by using the correct materials. Both these requirements will place additional burdens on the principal contractors to ensure that any subcontracts include the necessary provisions to ensure that sustainability and environmental issues are covered.

6.8.3 Land remediation

The RICS highlights that many sustainable building rating systems take into account land use. Valuers should take land use into account where buildings are constructed on brownfield land and watercourse setbacks. Where land has been constructed on previously developed land, issues of potential contamination must be considered as this may bear a risk of outlay in terms of cost and/or insurance against potential problems in the future.

It is not only valuers who should consider the potential risk of contamination. Developers intending to build property will need to carry out an environmental assessment of the site, and remove all environmental and health risks before construction can begin.

In the UK the Sustainable Remediation Forum UK (SuRF) has published a Framework for Assessing the Sustainable Remediation of Soil and Groundwater.[14] The document presents a framework for assessing the sustainable remediation of soil and groundwater remediation and incorporating the criteria of sustainable development into contaminated land management strategies. Its aim is to realize a number of benefits, including contributing to sustainable development at a number of levels and to positively demonstrate corporate and environmental responsibility.

6.9 Ownership and occupation of a building

6.9.1 Energy performance of buildings

The Energy Performance of Buildings (Certificates and Inspections) (England and Wales) Regulations 2007 require that the energy performance of completed building be assessed. All buildings that are bought, sold, rented or leased need an Energy Performance Certificate (EPC). An EPC has to be provided by the developer to the local authority building control, failing which the issuing of a certificate of practical completion may not be possible.

6.10 The sale or rental of a property

6.10.1 Valuation

When buying property, the issue is what price is or should be paid (either as a capital value or the rental value). In commercial property terms, the question that arises is whether these should reflect sustainability credentials.

The Royal Institution of Chartered Surveyors (RICS) is the professional body for valuers who are the main agents involved in assessing the value of buildings. The RICS is responsible for laying down guidelines for valuing assets for purposes such as secured lending or financial reporting. Valuers should, therefore, have access to reliable and timely information, including information on sustainability criteria if this is going to affect the valuation process. The RICS acknowledges that there are fundamental aspects of sustainability that affect property and potentially its value.

In September 2009 the RICS issued an Information Paper on Sustainability and Commercial Property Valuation.[15] This paper was drafted as a step towards embedding sustainability as a core consideration in the valuation process. The paper outlines ways in which sustainability can be defined, how a building's green credentials may be assessed, and how such characteristics might be reflected within a valuation – whether quantitatively or qualitatively. The Red Book was published in January 2010 with more substantial guidance on these issues.

The RICS guidance note on contamination, the environment and sustainability embodies 'best practice' and reminds members that whilst they are not required to follow the advice and recommendations, when an allegation of professional negligence is made against a surveyor, the court is likely to take account of the contents of any relevant guidance notes published by the RICS in deciding whether or not the surveyor has acted with reasonable competence.[16]

Valuation is key to investment and the provision of information is critical to ensuring that sustainability is built into investment decisions.

6.10.2 Energy performance certificates

The sellers and landlords of UK buildings are required by law to provide an EPC to potential buyers and tenants, setting out information on the building's energy efficiency. This is intended to allow potential buyers to make informed decisions on property investment based on a property's green credentials.

However, evidence suggests low levels of compliance. A monthly index, run by National Energy Services (NES) and Building.co.uk monitors how many commercial buildings currently being marketed have a valid EPC. The sample for February 2010 covered 1,084 buildings in Cumbria, Buckinghamshire, East Sussex and Leicester with a floor area in excess of $50m^2$ and included buildings which had been on the market for at least six months. Only 39 per cent of the properties investigated were compliant. Figures for June 2010 indicated a slight improvement to 44 per cent compliance.[17]

The index provides empirical support for concerns raised in the House of Lords by Lord Dixon-Smith who, in July 2009, referred to 'almost total ignorance or disregard of the need for energy performance certificates in the commercial sector'.[18]

Having lamented the lack of compliance, Lord Dixon-Smith described EPCs as 'the one really useful tool to come out of the Government's regulation of the property market'. Even allowing for damnation by faint praise, it is arguable that EPCs are widely ignored not only because of weak or inadequate enforcement, but also because their practical worth is extremely limited.

The energy performance rating in an EPC compares current energy efficiency and carbon dioxide emissions with potential figures a building could achieve. Potential figures are calculated by estimating what its energy efficiency and what carbon dioxide emissions would be if energy saving measures were put in place.

The certificate is accompanied by recommendations for energy saving measures. There is no legislative compulsion to make those improvements, and in recession-hit market conditions there has been very little incentive to do so. Tenants' solicitors were quick to point out that existing leases generally do not permit a landlord to insist on the tenant making or funding energy saving measures. While leases usually include a tenant's obligation to comply with statutory requirements relating to the property, there is no contractual obligation to comply with recommendations, even if they are made under statute.

Landlords have also been (rightly) concerned that a clause which unequivocally requires a tenant to permit and pay for energy improvements would be seized upon to argue for a discount at rent review.

In essence, EPCs for commercial property do not point towards or facilitate a practical outcome where the property is already occupied. That problem is amplified whenever tenants occupy business premises on leases protected by the Landlord and Tenant Act 1954. That Act gives tenants the right to renew on terms carried over from the previous lease. A landlord seeking to vary those terms must justify any variation (*O'May v City of London Real Property Co* [1982] 1 All ER 660). It would be extremely difficult to justify a new provision that departed from the underlying principle that tenants are required to pay for repairs, not for improvements.

Given that the majority of private sector commercial property is existing building stock, and that a high proportion is already occupied by tenants, there is a strong argument that EPCs could be made useful as a means of assessing a potential investment purchase or occupational lease only if they adopted the criteria used for public sector buildings.

A Display Energy Certificate (DEC) gives an operational rating for a large public building. It indicates the carbon dioxide emissions that result from the energy actually consumed over a period of twelve months, as recorded by gas, electricity and other meters. This allows a meaningful comparison to the benchmark, which is the typical quantity of carbon emissions for that particular type of building.

To enable the performance of one building to be compared with another, the government uses rules that take into account adjustable factors, such as:

- The location of a building.
- How many hours a building is occupied.
- Whether a building has mixed use.

Armed with an operational rating to measure against the asset (or potential) rating, a prospective buyer or tenant would have far more useful information to assess the relative merits of similar buildings. For landlords subject to the CRC Energy Efficiency Scheme, that information might significantly inform the choice of investment, identifying properties where significant improvements could be achieved by incentivizing tenant behaviour rather than by making expensive alterations. Ultimately, compliance is more likely to be driven by economics than by regulatory burden.

6.10.3 Green leases

The adoption of lease arrangements that either encourage or require landlords and tenants to manage the asset in accordance with sustainability principles (so called green leases) is growing. The RICS recommends in its Valuation Information Paper No 13[19] that the presence of a green lease is to be evaluated as it could mark a risk reduction factor within the appraisal or, conversely, it could result in a lower rental bid if it contains onerous terms.

In the UK the Better Buildings Partnership (BBP)[20] provides a green lease toolkit that aims to improve sustainability and reduce the environmental footprint of existing buildings.

Major UK commercial landlords including Land Securities, British Land and Segro have sought to introduce green lease provisions into new lettings, and some attempts have been made by those landlords to add similar provisions to existing leases by means of a 'memorandum of understanding'. However, market factors and legislation affecting existing business premises create significant inertia in the sector when it comes to introducing energy performance improvements. Key factors include:

1   The need for agreement between landlord and tenant for any variation to an existing lease during its term. In particular, a tenant faced with a request for a variation that would increase its costs of occupation (e.g. seeking a contribution to costs incurred in improving the fabric, plant or equipment of premises, or requiring the tenant to meet the costs of CRC allowances) would require a significant incentive. The tenant is under no obligation to agree to a variation to the lease or to the addition of any side agreement or 'memorandum of understanding' importing CRC provisions.
2   Where a lease is protected by Part II Landlord and Tenant Act 1954, the tenant is generally entitled to a renewal on terms derived from the original lease. A party seeking to vary those terms in a renewal lease must demonstrate that the variations are reasonably required to modernize or update the lease. Tenants are likely to resist a variation that would alter the balance of responsibility for capital expenditure, or that would require them to fund improvements, citing the House of Lords ruling in *O'May v City of London Real Property Co* [1982] 1 All ER 660.

If tenants will not agree to variations, there is little or no incentive for landlords to incur expenditure on improvements. Indeed, during the term of a lease the tenant enjoys exclusive possession of premises, so the landlord may be unable to gain access to carry out any works. Consequently, penetration of energy efficiency measures in the commercial sector is likely to remain limited, and industry commentators are beginning to ask what could government do to encourage regearing?

The capital value of a commercial building is determined by the income stream from rents, the 'covenant strength' of the tenants and the remaining term of the leases. As a lease approaches its contractual expiry, with no certainty of renewal, capital value can drop sharply. Consequently, landlords frequently seek to 'regear' their leases. That process involves early renewal of the lease so that (for example) a term with less than three years remaining is converted into a new lease for ten or fifteen years. The positive impact on capital value can be significant. Where regearing produces a significant improvement for the landlord in terms of capital value, there is scope for sharing that benefit to induce the tenant to agree to the new terms, whether by way of a rent-free period or a contribution to works that the tenant wishes to carry out. If that benefit is significant enough, regearing would offer a major opportunity to carry out energy efficiency improvements and/ or to introduce 'green lease' or CRC provisions.

However, regearing schemes are inhibited by existing legislative and common law consequences including:

1   The replacement of an existing lease with one for a longer term always takes effect as a 'surrender and regrant' (*Friends Provident Life Office v British Railways Board* [1996] 1 All ER 336). Stamp duty land tax is payable on the new lease term. Consequently, in order to at least defer that liability, tenants may prefer to delay any decision on lease renewal until the contractual expiry date (or indeed until after that date unless the landlord triggers the statutory renewal or termination procedure under the Landlord and Tenant Act 1954. If that step is not taken, the existing lease is continued indefinitely by s 24 of that Act).
2   If the tenant agrees to regearing, and to the carrying out of improvement works, it may need to find temporary alternative premises, potentially taking on liability for business rates in respect of both premises.
3   For the landlord, a risk of regearing is the loss of existing covenants as guarantors and former tenants would be released by the 'surrender and regrant'. Often, the covenant strength of the guarantor is far greater than that of the tenant, so the prospect of losing the guarantor can militate strongly against regearing.

Regearing offers perhaps the best opportunity to effect energy efficiency improvements, to introduce 'green lease' provisions and to allocate CRC responsibilities in existing landlord and tenant relationships. Without it, penetration is likely to remain limited to new builds and new lettings. To remove the barriers to widespread regearing (and with it, to a significant strengthening or enhancement of capital values) government might consider:

1   A specific stamp duty land tax (SDLT) relief on regearing, where the parties can demonstrate that the new lease meets specified criteria, such as an improvement in the asset and/or operational rating of the building verified by inspection under the EPC/DEC regime, or the carrying out of works such as replacing inefficient plant and equipment recommended as part of the EPC/DEC process. Making SDLT relief conditional on such improvements would remove a significant barrier to regearing, and incentivize efficiency gains.

2   A specific business rates relief for any period (subject to a maximum to avoid abuse) during which the tenant is unable to occupy premises and/or must occupy alternative premises while improvement works are carried out in accordance with criteria replicating those for SDLT relief.

3   To meet landlords' concerns, a statutory provision continuing the obligations of any guarantor or former tenant until they would have ended had there been no 'surrender and regrant'. This would simply preserve the covenant strength available to the landlord for the duration that was originally agreed. It would not worsen the position of the guarantor or former tenant.

6.11 Carbon Reduction Commitment (CRC) energy efficiency scheme

The picture for commercial property is somewhat different from residential with the UK government using the Carbon Reduction Commitment (CRC) energy efficiency scheme to try and improve energy efficiency in this sector.

The CRC is a mandatory emissions trading scheme for the UK, designed to reduce carbon dioxide emissions of UK properties in both the commercial and public sector. The CRC commenced in April 2010 and will run for three set time periods (known as phases). The first phase will run for three years and the subsequent phases will each last for seven years. Poor compliance can potentially result in increased costs, bad publicity and criminal prosecution.

The CRC will have a significant impact on landlords, tenants and the investment market and the RICS recognizes that its impact has to be fully appraised and reference to the CRC will be included in subsequent editions of their guidance notes.

However, CRC is an intensely bureaucratic process. New lease clauses to address CRC empower landlords to pass on the cost of allowances and compliance to tenants by allocating costs to buildings within its portfolio on a 'fair and reasonable basis'. A calculation by reference to floor area might be ostensibly fair and simple to operate, but does not distinguish between types of building or intensity of energy use. Tenant amendments may require the landlord take into account the 'energy efficiency' of the relevant building as compared with any other building for which the landlord must purchase allowances. However, 'energy efficiency' requires further definition. It might relate to the 'asset rating' evidenced by an Energy Performance Certificate or to the 'operational efficiency' assessed by comparing an assumed level of consumption with actual usage. A tenant has no influence or control over the asset rating, but a great deal over the operational rating.

If energy use exceeds initial allowances, a landlord seeking to pass on to tenants the cost of additional allowances might expect strenuous objections from those whose energy use remained within the original allowances, arguing that the additional costs should be borne by profligate tenants. Identifying those tenants might be extremely difficult in a large portfolio, generating significant administrative cost and inconvenience in meeting tenant demands for information.

Commercial activity on the part of the landlord might itself necessitate additional allowances. Existing tenants would be likely to resist demand for further payments triggered by the landlord's commercial decision, while tenants of the newly acquired property might find themselves for the first time dealing with a landlord subject to CRC and demanding previously unbudgeted payments.

Recycling payments were intended to incentivize CRC entities and improve energy efficiency. If landlords required tenants to pay the up-front cost of allowances they could have no reasonable basis for retaining recycling payments for their own benefit. However, in the hands of tenants, recycling payments would have been highly unlikely to meet the policy objectives of CRC, and would certainly not have assisted landlords to meet the demands of transition to a low carbon economy. Tenants, particularly those with relatively short terms, would have absolutely no obligation or incentive to use those payments to fund improvements to the landlord's capital asset.

By contrast, in the hands of the landlord, recycling payments might have been allied with a range of incentives and penalties and legitimate service charge recovery to form part of a fund for the progressive renewal or improvement of plant, equipment and energy performance.

Faced with new costs, many landlords will adopt a 'business as usual' approach and seek to pass them on to tenants. However, the British Property Federation's acknowledgement in June 2010 that no common ground could be found between landlords and tenants in the quest for 'standard' CRC lease clauses underlines the point that landlords cannot assume that tenants will meekly pick up the bill. CRC is a new and different type of cost, and calls for a rethink of traditional institutional assumptions.

6.12 CRC – the UK's new 'carbon tax'?

On 20 October 2010 the Chancellor announced that money raised from the sale of allowances under the CRC energy efficiency scheme will be diverted to the Treasury and 'used to support the public finances (including spending on the environment), rather than recycled to participants'. The announcement marks a radical departure from the scheme which, during extensive consultation, was presented as 'revenue neutral' rather than 'revenue raising'. The CBI immediately denounced the change of plan as a 'stealth tax'. In fact, there is no stealth. Once implemented, CRC will have all the hallmarks of a tax, reopening the acrimonious debate between landlords and tenants over who should pay.

The announcement is bad news for tenants who have accepted specific obligations to meet the costs of allowances and administration. In many cases, tenant resistance to such clauses has been overcome by amendments promising reimbursement of the whole or a fair proportion of 'revenue recycling' payments 'received by' or 'due to' the landlord. In practice, those amendments were always vulnerable. CRC operates at corporate group level. The CRC participant might be a parent company several rungs up the landlord's corporate ladder. The landlord company might have received and, in the absence of specific contractual arrangements within the group, have been entitled to nothing. With that risk in mind, some tenants took the logical step of requiring the landlord either to pay or to 'procure' payments. However, with funds going to the Treasury, even that approach is now worthless. Tenants are left with specific obligations to pay, and no mitigation through revenue recycling.

6.12.1 A Pyrrhic victory?

Arguably, the news is even worse for tenants whose lease obligations predate CRC, and for those who successfully resisted CRC obligations in leases negotiated during the past year. Landlords' perceived need for CRC clauses stemmed in large part from a concern that costs could not be passed on using existing lease provisions. In particular, it was considered unlikely that costs could be passed on using the tenant's general covenant to pay all taxes and outgoings relating to its use and occupation of the premises.

The diversion of CRC funds to the Treasury mean that the scheme now looks very much like a tax, and successfully resisting specific CRC clauses may prove to have been a Pyrrhic victory.

The House of Lords considered the hallmarks of taxation in *Aston Cantlow v Wallbank* [2003] 3 All ER 1213. Lord Scott approved the description of 'a charge by the government . . . a pecuniary burden laid upon individuals or property to support the government, exacted by legislative authority'. It is an enforced contribution enacted pursuant to legislative authority. This is entirely consistent with other commonwealth jurisdictions (e.g. *Inland Revenue Commissioner and Attorney General v Lilleyman* (1964) 7 W.I.R 496; *Smith v Ministry of Housing and National Insurance* [1988] BHS J. No. 90). Converted from 'revenue neutral' to 'revenue raising' status, CRC seems to tick all of the relevant boxes.

It is highly likely that some hard-pressed landlords will now argue that existing lease obligations can be used to pass costs on to their tenants. Their likely magnitude certainly makes the argument worthwhile. KPMG estimates that scrapping revenue recycling could represent a five- to ten-fold increase in the costs of compliance. Penalties remain, but without the incentivizing effect of weighted repayments, based on relative performance.

6.12.2 A tax on use or occupation?

Faced with a demand for payment, tenants might argue that the reconfigured CRC does not fall within the general obligation to pay taxes and outgoings contained in most leases. CRC is a tax levied on the highest UK parent company in a group. The requirement to register as a CRC participant, and the extent of liability, are determined by the energy in respect of which group members are direct contracting parties. The tenant of multi-let premises might observe that the CRC participant status of the landlord's group could be determined by energy consumed at its owner-occupied corporate headquarters or in other, more energy-intensive, parts of its portfolio. However, landlords would no doubt counter by arguing that the extent of CRC liability is determined by energy use, and that in turn depends on occupiers' behaviour. From a landlord's perspective, the causal connection between occupiers' energy consumption and liability is likely to provide a sufficient basis for demanding tenant payments.

The incentive to press for payment is amplified where landlords have invested in Carbon Trust Standard certification and other Early Action Measures. Without revenue recycling, the benefits will be severely diluted and return on investment can be salvaged only by minimizing compliance costs and applying financial pressure on tenants to prompt energy-saving behaviour.

6.12.3 Open to challenge?

The coalition government's withdrawal of revenue recycling payments departs from the scheme that was the subject of extensive consultation, based on regulatory impact assessments stressing its 'revenue neutral' status. The result is a new tax, with arbitrary thresholds and with no clear current basis in primary legislation. The Climate Change Act 2008 mandated the creation of an emissions trading scheme, not a new carbon tax. Meanwhile, larger greenhouse gas emitters, covered by the EU Emissions Trading Scheme, continue to receive free emissions allowances. It will be interesting to see whether the legislation giving effect to the CRC announcement is sufficient to head off any tenant challenge to the new tax as discriminatory and contrary to Article 14 of the ECHR.

6.13 Alteration and retrofit

The alteration or extension of a commercial building in England and Wales is governed by the Building Regulations 2000. Any building work will need to meet minimum sustainability standards covering energy efficiency and water (amongst others). If any alterations or retrofits are being carried out, the developers will have to consider whether planning permission is required.

Building materials such as cladding, ceiling and floor panels, and carpets and walling are important sustainability considerations. The life cycle value of materials should be considered as part of the rental value of a building.

Where retrofitting includes an element of renewable energy generation, commercial and large-scale residential landlords must also consider the interaction between the 'feed-in tariffs' regime introduced in April 2010 and its obligations under CRC. The regimes are mutually exclusive. A landlord might elect to install renewable energy apparatus (such as solar photovoltaic panels) and to receive 'feed-in tariff' payments for the electricity generated, with a higher rate for energy exported to the grid. While a useful income stream for landlords, power for which feed-in tariffs are claimed must be included in CRC returns, and must be covered by CRC allowances. Consequently, tenants who accept an obligation to meet the cost of CRC allowances could find themselves paying for energy in respect of which the landlord has received payment.

Faced with that prospect, tenants might seek to insist that landlords elect to take CRC electricity-generating credits instead of feed in tariff payments. CRC electricity-generating credits reduce the energy that must be reported and covered by allowances under CRC.

For many landlords, there will be a strong case in favour of opting for feed-in tariffs rather than CRC electricity-generating credits. Retrofitting a building with renewable energy apparatus offers the prospect of a new income stream from otherwise under-utilized parts of a building, in some cases replacing income lost due to consolidation of electronic communications networks and the consequent reduction in telecommunications base station sites. Further, where an installation produces a surplus of energy, the lure of higher rates of feed-in tariff for exports to the grid may well be determinative.

6.14 Demolition

Planning permission is required prior to demolishing a building. Developers will also have to comply with the Site Waste Management Regulations 2008 and BR2000 throughout the demolition process.

6.15 Challenges for businesses

The real estate sector can no longer afford to ignore the movement to establish a low-carbon sustainable global economy. Corporate bottom lines will be impacted by emerging ESG issues, particularly where disclosure must now be included in financial filings. Impacts will flow from the physical risks arising from climate change (whether or not anthropogenic) or the physical and economic risks arising from the scarcity of energy and water resources, or from legal and regulatory burdens which may be imposed by reference to those risks. The overall governance and disclosure by companies relating to ESG issues must be improved. More details are required on actions and policies on assessing and managing environmental risks.

Companies within the real estate sector should also be aware of the growing trend of the global business and human rights agenda. Companies are also going to have to track their policies and operations on the local communities within which they operate.[21]

Companies can also not ignore the fact that new social technologies, media and networks will transform the reporting landscape. The Web 2.0 revolution is levelling the playing field by giving stakeholders the opportunity to initiate and drive conversations with (or around) companies (GRI, 2010).

The risks of inaction by property investors, failing to embrace energy efficiency, are indentified in the Ceres report as:

- Expected rises in energy costs, including the cost of waste consumption
- Existing and new, more stringent legislation
- Competitive and financial risks of not responding to market demand

Those companies embracing the sustainability agenda are likely to see decreased operating costs of certified buildings and combined with increased rental levels, the cash flows of the sustainable buildings should be more attractive.

Those investors and businesses pursuing energy efficiency initiatives should position them competitively for impending national climate and energy legislation that will likely make energy consumption and waste more expensive.

Global real estate businesses, with an international property portfolio, are faced with the challenge that there are inconsistencies between currently available metrics. For example, the EPBD implementation mechanism varies across member states, hence buildings cannot be compared for energy efficiency across Europe in a consistent manner.

The impact of climate change on the ability to continue to use a building efficiently in high temperatures or withstanding storms will be important. Many properties could become unsuitable in extreme heat without appropriate climate control which will impact on tenants. Such buildings may be vulnerable to obsolescence and may need retrofitting.

For those architects and developers looking to design and build new developments, there is going to be an increasing need to be aware of the current and future legal requirements and voluntary standards which may affect the design of the development. Developers will also have to liaise with the local planning authority to understand the planning policy requirements and budget for increased costs arising from compliance with sustainability standards.

Commercial organizations that may qualify under the CRC scheme (if they have not already done so) should consider how they can resource and implement new systems to capture data and report it, in order to be compliant and also benefit from the league table ranking that will predominately be based on the first two CRC compliance years.

Commercial owners and occupiers must also consider the appropriate balance between their obligations under CRC and the opportunities afforded by initiatives such as 'feed-in tariffs' (FITs) which came into operation in April 2010. FITs provide payments for 'microgeneration' installations such as solar photovoltaic panels, wind turbines or anaerobic digestion equipment. The regime also provides for electricity to be exported to the grid with an 'obligation to buy' being placed on energy supply licence holders using powers conferred by Energy Act 2008, s 41.

FITs and CRC are mutually exclusive regimes. If FITs are claimed, CRC electricity generating credits are not available, and energy must be accounted for and covered by CRC allowances. Consequently, a commercial landlord's decision to take FITs or a rental income derived from FITs, and to retain that income for its own benefit, would potentially disadvantage tenants whose obligations included the funding of CRC allowances. A significant imbalance might weaken the argument for a 'premium' rent from such tenants.

The challenge for the property market will be the creation and adoption of standardized approaches to sustainability, from the valuation process to the design and construction of buildings. The occupation of buildings will also be affected by the legislative frameworks and market pressures. Creation of a universal definition of what constitutes a sustainable building and simple benchmarks will be vital.

Notes

1   Available at: http://www.lexisnexis.com/Community/environmental-climatechange law/landing/EmergingIssues.aspx.
2   Available at: http://www.guardian.co.uk/environment/2010/may/12/coalition-environment-policy.
3   Available at: http://www.eres2009.com/papers/1A_Schleich.pdf.
4   Brundtland Report (1987): http://www.un-documents.net/wced-ocf.htm.
5   Ceres is a national coalition of investors, environmental groups and other public interested organizations working with companies to address sustainability challenges such as water scarcity and climate change. See: http://www.ceres.org.
6   Available at: http://www.rics.org/site/download_feed.aspx?fileID=5752&file Extension=PDF.
7   Available at: http://www.joneslanglasalle.com/ResearchLevel1/Global_Trends_ in_Sustainable_Real_Estate_-_Feb_2008_EN.pdf.
8   Available at: http://www.kingsturge.com/en-gb/international-research/%7E/media/ F0708881EF2742269586BEC98A17D2C7.ashx.
9   Available at: http://www.ceres.org/Document.Doc?id=519.
10 Mercer is a global consultancy service.
11 This assessment was launched in June 2009 by AXA Real Estate Investment and ING Real Estate amongst others.
12 Available at: http://www.communities.gov.uk/documents/corporate/pdf/163592 12.pdf.

13 Available at: http://www.jctltd.co.uk/assets/Building%20a%20sustainable%20 future%20together%202009%20Web.pdf.
14 Available at: http://www.claire.co.uk/index.php?option=com_content&task=view &id=182&Itemid=78.
15 See http://www.rics.org/site/download_feed.aspx?fileID=5751&fileExtension=PDF.
16 RICS Practice Standards, UK; page 01.
17 Available at: http://www.nesltd.co.uk/news/commercial-epc-compliance-44.
18 HL Deb 2 July 2009, c 323.
19 Available through the RICS membership.
20 Available at: http://www.bitc.org.uk/resources/publications/green_lease_toolkit. html.
21 http://www.euractiv.com/en/print/socialeurope/csr-corporate-social-responsibility/ article-153515.

References

Ceres (January 2010) 'Investors Analyze Climate Risks and Opportunities: A survey of Asset Managers' Practices' http://www.ceres.org/Document.Doc?id=519.

Cushman and Wakefield (2009) 'Landlord & Tenant Survey', London.

Loew, T., Ankele, K., Braun, S., Clausen, J. (2004) 'Significance of the CSR Debate for Sustainability and the Requirements for Companies' http://www.csr-weltweit. de/uploads/tx_jpdownloads/future-IOEW_Significance_CSR_debate_sustainability. pdf.

Jones Lang LaSalle (2008) 'Global Trends in Sustainable Real Estate: An Occupier's Perspective' http://www.joneslanglasalle.com/ResearchLevel1/Global_Trends_in_ Sustainable_Real_Estate_-_Feb_2008_EN.pdf.

Keay, A. (1990) 'Insolvency and Environmental Principles: A Case Study in Conflict of Public Interests' *Environmental Law Review*, Volume 3 (2).

Miller, N., Spivey, J., Florance, A. (2008) 'Does Green Pay Off?' *Journal of Real Estate Portfolio Management*, Volume 1 (4), 385–399.

Schleich, H., Lindholm, A. and Flakenback, H. (2009) 'Environmental Sustainability-drivers for the Real Estate Investor' http://www.eres2009.com/papers/1A_Schleich. pdf.

The GRI Learning Series (2010) 'The Transparent Economy', p. 15.

Part III

7 Sustainable development and the South African Constitution: implications for built environment legislation

Jeremy Gibberd

7.1 Introduction

South Africa faces a range of social, economic and environmental challenges. HIV/Aids has resulted in life expectancy dropping from 67 years in 1998, to approximately 47 years and unemployment is estimated to be 27 per cent (2009 figures). South Africa also has one of the highest CO_2 emissions per GDP in the world and in 2002 carbon emissions per capita in South Africa were 8.4 tonnes/capita – higher than Western European averages of 7.9 tonnes/capita (Sustainable Energy Africa, 2006).

Global warming will make South Africa's social, economic and environmental problems worse. The National Climate Change Response Policy developed by South Africa's Department of Environment and Tourism outlines the following impacts of climate change on Africa (Department of Environment and Tourism, 2009):

- Agricultural production and food security in many African countries are likely to be severely compromised by climate change and variability. Projected yields in some countries may be reduced by as much as 50 per cent by 2020 and as much as 100 per cent by 2100. Small-scale farmers will be most severely affected.
- Existing water stresses will be aggravated. About 25 per cent of Africa's population (about 200 million people) currently experience high water stress. This is projected to increase to 75–250 million by 2020 and 350–600 million by 2050.
- Changes in ecosystems are already being detected and the proportion of arid and semi-arid lands in Africa is likely to increase by 5–8 per cent by 2080. It is projected that between 25 and 40 per cent of mammal species in national parks in sub-Saharan Africa will become endangered.
- Projected sea-level rises will have implications for human health and the physical vulnerability of coastal cities. The cost of adaptation to sea level rise could amount to 5–10 per cent of gross domestic product.
- Human health will be negatively affected by climate change and vulnerability and incidences of malaria, dengue fever, meningitis and cholera may increase.

The scale and breadth of the challenges facing South Africa mean that it is imperative to have legislation that will address these environmental, social and economic problems. It is therefore argued that the South African Constitution and building-related legislation must provide an appropriate legal framework to ensure that sustainable development is effectively integrated into buildings and construction.

7.2 The South African Constitution

The South African Constitution was developed in 1996. As the supreme law of South Africa it may not be superseded, and government and other parties may not violate provisions within this (South African Constitutional Court 2009). It contains a Bill of Rights that enshrines the rights of all people in South Africa and affirms the democratic values of human dignity, equality and freedom. The Bill has sections covering equality, human dignity, privacy, freedom of religious belief and opinion, environment, property, housing, healthcare, food, water and social security, children, education, language and culture. Through a section on equality, the Bill requires that all people have full and equal enjoyment of these rights and freedoms:

> Everyone is equal before the law and has the right to equal protection and benefit of the law.

Equality includes the full and equal enjoyment of all rights and freedoms. To promote the achievement of equality, legislative and other measures designed to protect or advance persons, or categories of persons, disadvantaged by unfair discrimination may be taken.[1]

Rights in the Bill are, however, subject to limitations. These are outlined in Section 36 of the Bill:

> The rights in the Bill of Rights may be limited only in terms of law of general application to the extent that the limitation is reasonable and justifiable in an open and democratic society based on human dignity, equality and freedom, taking into account all relevant factors, including
>
> a. the nature of the right;
> b. the importance of the purpose of the limitation;
> c. the nature and extent of the limitation;
> d. the relation between the limitation and its purpose; and
> e. less restrictive means to achieve the purpose.[2]

The role and responsibility of government is also outlined in the Bill. This specifically requires government to achieve the rights outlined in the Bill:

> The state must respect, protect, promote and fulfil the rights in the Bill of Rights.[3]

Environmental rights in the Bill of Rights include the right to an environment that supports health and well-being. It also requires legislation to be developed to ensure that the environment is protected, and that development is both sustainable and justifiable:

24. Environment

Everyone has the right

a. to an environment that is not harmful to their health or well-being; and
b. to have the environment protected, for the benefit of present and future generations, through reasonable legislative and other measures that
   i. prevent pollution and ecological degradation;
   ii. promote conservation; and
   iii. secure ecologically sustainable development and use of natural resources while promoting justifiable economic and social development[4]

7.2.1 Implications of Section 24 of the Constitution for the built environment

There is a range of implications for the built environment of Section 24 from the Bill of Rights, outlined above. This requires some interpretation and for the following questions to be answered:

- What is defined as an environment?
- How is health and well-being defined?
- What does ecologically sustainable development mean?
- What is meant by justifiable economic and social development?

7.3 The environment

There is no comprehensive international treaty on human rights and environment that can be used to understand how the environment should be defined in the Bill of Rights (South African Human Rights Commission, 1996a). The 1972 Stockholm Declaration on the Human Environment adopted by the UN Conference on the Human Environment, however, begins to define rights and obligations of man in regard to the environment:

Man has the fundamental right to freedom, equality and adequate conditions of life, in an environment of quality that permits a life of dignity and well being, and he bears the solemn responsibility to protect and improve the environment for present and future[5]

The African Charter on Human and People's Rights adopted by heads of states of the Organisation for African Unity in 1981 also sets out how the environment should be interpreted in an African context:

> All peoples shall have the right to a general satisfactory environment favourable to their development.[6]

Section 24 of the Bill of Rights from the South African Constitution indicates that the environment can be defined both as a natural and man-made environment. De Waal suggests that this definition includes 'man made objects and cultural and historical heritage' (De Waal *et al.*, 1999). Feris argues therefore that in interpreting Section 24, traditional rights, needs and values and the dignity of indigenous people should be taken into account (South African Human Rights Commission, 1996b).

7.3.1 Implications for the built environment

Therefore, the following implications of the South African Constitution can be inferred for the built environment:

* The built environment should create environments favourable for the development of people.
* Built environments should take into account the traditional rights, needs and values of indigenous peoples.

7.4 Health and well-being

The World Health Organization (WHO) defines health in the following way:

> Health is a state of complete physical, mental and social well-being and not merely the absence of disease or infirmity.[7]

Given this broad definition it is clear that environments conducive to health and well-being are likely to be described in different ways by different people and socio-economic groups. Wealthy people may wish to protect the environment in order to avoid mental or aesthetic discomfort. Poor rural people, on the other hand, would want to protect the environment because they rely on this for clean water and food. (South African Human Rights Commission, 1996c).

7.4.1 Implications for the built environment

Therefore, the following implications of the South African Constitution can be inferred for the built environment:

- The built environment should support physical, mental and social well being.
- Criteria used to define physical, mental and social well-being in built environment must take into account, and respond to, the particular situation of the people who occupy and use these built environments.

7.5 Ecologically sustainable development

Principles from the Rio Declaration on the Environment and Development can be used to define sustainable development. The principles below from the Declaration indicate that development must be equitable and that environmental protection must be integrated into development:

Principle 3
The right to development must be fulfilled so as to equitably meet developmental and environmental needs of present and future generations.

Principle 4
In order to achieve sustainable development, environmental protection shall constitute an integral part of the development process and cannot be considered in isolation from it.[8]

7.5.1 Implications for the built environment

The following implications of the definition for sustainable development outlined above for the built environment can be inferred:

- Development of the built environment should ensure that developmental and environmental needs of present and future generations are achieved in an equitable way.
- Conservation and protection of the environment should be integrated into any built environment development process.

7.6 Justifiable economic and social development

The meaning of 'justifiable economic and social development' found in Section 24 of the Bill of Rights can be interpreted through reference to other sections of the Bill of Rights. These sections include rights to housing, healthcare, food, water and social security and education:

26. Housing
Everyone has the right to have access to adequate housing.[9]

27. Health care, food, water and social security
Everyone has the right to have access to
a. health care services, including reproductive health care;[10]

29. Education

Everyone has the right

b. to a basic education, including adult basic education; and

c. to further education, which the state, through reasonable measures, must make progressively available and accessible.

d. sufficient food and water;[11]

It can be argued that development that helps to ensure that these rights are achieved can be classified as *justifiable economic and social development*. Conversely, development that does not directly contribute to the achievement of these rights may be deemed to be less justifiable.

7.6.1 Implications for the built environment

Therefore the following implications of the South African Constitution can be inferred for the built environment:

* Development of built environments that directly contribute to the achievement of rights in the Bill of Rights, may be deemed to be justifiable.
* Development of built environments that do not directly contribute to the rights included in the Bill of Rights may be deemed to be not justifiable.

7.7 Built environment legislation

A review of the South African Constitution shows that sustainable development is an explicit requirement. It also shows that sustainable development, as defined in the Constitution, encompasses social, economic and environmental development and that clear objectives for the built environment can be inferred.

   Given these objectives it is possible to review current building-related legislation in order to ascertain whether this is effective in upholding the Constitution and sustainable development. Key legislation used to control building development in South Africa includes the Building Regulations (South African Bureau of Standards 1990), the Occupational Safety and Health Act (Department of Labour 1987, 1988, 2003), and the National Environmental Management Act (Department of Environment and Tourism 1998). This legislation is reviewed below.

7.8 Health and well-being

Legislation on health and well-being in built environments can be found in the Building Regulations and in the Occupational Health and Safety Act. The Building Regulations set out minimum requirements for lighting and ventilation (South African Bureau of Standards, 1990a). Other aspects such as thermal and acoustic performance are not dealt with because, it is argued, these can only be judged in a subjective manner:

There are other aspects to a building which may affect only the comfort or convenience of people but many of these, such as acoustic or thermal performance, are judged in a subjective way and are not readily amenable to control in a sensible manner by regulation. It is also obvious that the market will limit the degree to which these matters can be considered in the design of a building.

The Occupational Health and Safety Act addresses health and well-being through the Facilities Regulations and the Environmental Regulations for Workplaces. The Facilities Regulations require minimum standards for sanitation, changing rooms, dining rooms, drinking water and seating. The Environment Regulations for Workplaces set out requirements for hot and cold working environments, lighting, windows, ventilation, space and noise.

The legislation does not address thermal conditions in working or living environments where temperatures are between 0 and 30°C. Temperatures between this range, for instance, 2°C and 29°C are therefore deemed acceptable, even though these may not be conducive to health and well-being. For instance, human comfort is indicated as being between 20°C and 27°C on the comfort chart used by the American Society of Heating, Refrigerating and Air-Conditioning Engineers, 1992 (American Society of Heating, Refrigerating and Air-Conditioning Engineers, 1992). This gap in legislation has resulted in many buildings such as housing, offices and classrooms being built without insulation or cooling or heating systems.

The legislation is also not responsive to different requirements of sections of the population in terms of health and well-being. For instance, babies, children and sick people who are more sensitive to heat and cold are not catered for, and temperatures in school environments can reach 45°C (Dolley and Hermanus, 2009).

Similarly, legislation allows poorly constructed housing with no insulation to be built. This is often occupied by people with few resources to counter, through heating or cooling, discomfort and ill health caused by high or low temperatures (Mathews *et al.*, 1995). Open fires used to heat accommodation have led to suspended particulates being found to be from 3 to 12 times higher than those prescribed by the World Health Organization (Terblanche, 1992).

It can therefore be argued that building-related legislation does not sufficiently uphold the right to health and well-being outlined in Section 24 of the Bill of Rights. In particular, building legislation does not prescribe any minimum thermal or acoustic standards in building. The lack of legislation in this area can lead to thermal and acoustic environments in buildings which are damaging to health and well-being. In addition, the legislation does not address other issues such as the presence of volatile organic compounds in air as a result of off-gassing from carpets, adhesive or paint which have also been shown to have a harmful effect on health (Wieslander *et al.*, 1996).

7.9 Indigenous construction

The South African building regulations are performance based and aim to avoid prescriptive requirements (South African Bureau of Standards, 1990b). Theoretically this approach should allow, and encourage, alternative and indigenous construction materials and designs that are more affordable and supportive of health and well-being. In reality, however, local authorities have been wary of approving anything considered 'alternative construction' without further information, as outlined below (South African Bureau of Standards, 1990c).

> Where there is doubt as to the efficiency of any design or method of construction proposed, the local authority may call for further information which normally would take the form of one or more of the following –
> (i)   a test report from the SABS;[12]
> (ii)   a test report from the CSIR;[13]
> (iii) an Agrément Certificate;[14]
> (iv) verification of a design by an independent Professional Engineer.

This clause can be used by local authorities to prevent any construction other than the standard 'European'-type construction. This has resulted in most buildings in South African towns being constructed of brick and corrugated steel, even though in rural, unregulated areas, mud brick and thatch are traditionally used as building materials.

It therefore could be argued that building-related legislation does not sufficiently uphold traditional rights, needs and values and the dignity of indigenous people, if this is included in the definition of 'well-being' outlined in Section 24 of the Constitution. It does this by making it more difficult, and more expensive, for alternative and indigenous building methods to be used by requiring test reports, certificates or verification from a professional engineer.

7.10 Ecologically sustainable development

The building regulations and the Occupation Health and Safety Act do not include any reference to sustainable development or conservation and protection of the environment. This, however, is addressed by the National Environment and Management Act (NEMA) (Department of Environment and Tourism 1998). This states that:

> (3) Development must be socially, environmentally and economically sustainable.
> (4) (a) Sustainable development requires the consideration of all relevant factors including the following:
> i.   That the disturbance of ecosystems and loss of biological diversity are avoided, or, where they cannot be altogether avoided, are minimised and remedied;

ii. that pollution and degradation of the environment are avoided, or, where they cannot be altogether avoided, are minimised and remedied;

iii. that the disturbance of landscapes and sites that constitute the nation's cultural heritage is avoided, or where it cannot be altogether avoided, is minimised and remedied;

iv. that waste is avoided, or where it cannot be altogether avoided, minimised and reused or recycled where possible and otherwise disposed of in a responsible manner;

v. that the use and exploitation of non renewable natural resources is responsible and equitable, and takes into account the consequences of the depletion of the resource;

vi. that the development, use and exploitation of renewable resources and the ecosystems of which they are part do not exceed the level beyond which their integrity is jeopardised;

vii. that a risk averse and cautious approach is applied, which takes into account the limits of current knowledge about the consequences of decisions and actions; and

viii. that negative impacts on the environment and on people's environmental rights be anticipated and prevented, and where they cannot be altogether prevented, are minimised and remedied.[15]

In order to implement these principles NEMA uses a number of mechanisms including the application of environmental management tools. These are required to (Department of Environment and Tourism, 1998):

a. promote the integration of the principles of environmental management set out in section 2 into the making of all decisions which may have a significant effect on the environment;

b. identify, predict and evaluate the actual and potential impact on the environment, socioeconomic conditions and cultural heritage, the risks and consequences and alternatives and options for mitigation of activities, with a view to minimising negative impacts, maximising benefits, and promoting compliance with the principles of environmental management set out in section 2;

c. ensure that the effects of activities on the environment receive adequate consideration before actions are taken in connection with them;

d. ensure adequate and appropriate opportunity for public participation in decisions that may affect the environment;

e. ensure the consideration of environmental attributes in management and decision making which may have a significant effect on the environment; and

f. identify and employ the modes of environmental management best suited to ensuring that a particular activity is pursued in accordance with the principles of environmental management set out in section 2.[16]

In addition, NEMA allows the minister to identify activities which may not be commenced without prior authorization. It can also be used to identify areas in which specified activities may not be commenced without prior authorization (Department of Environment and Tourism, 1998):

(a) identify activities which may not be commenced without prior authorisation from the Minister or MEC;
(b) identify geographical areas in which specified activities may not be commenced without prior authorisation from the Minister or MEC and specify such activities[17]

While this legislation is a clear attempt to support Section 24 of the constitution, there is a gap between the principles espoused in the Act and how this is implemented. Chapter 1 of NEMA (see above) states that development must be socially, environmentally and economically sustainable. This, however, is never fully defined or followed through in the Act, which instead dwells at some length on defining development activities and geographical areas where legislative control may be exerted. The legislation is not explicit on what types of development would be considered 'more' or 'less' sustainable. Thus the Act does not prescribe, or give specific preference to, for instance, developments such as high-density social housing which may be seen as more sustainable relative to developments such as golf courses or casinos which may be seen as less sustainable. In addition, the Act does not list or provide criteria which can be used to establish whether a development is '*socially, environmentally and economically sustainable*'.

It can therefore be argued that current legislation does not adequately define or describe sustainable development, or ensure that this is achieved. Existing legislation tends to be used to control particular types of development, in specified geographical areas. It can be argued that sustainable development, as defined in the constitution, has broader and more far-reaching implications.

7.11 Justifiable economic and social development

Building-related legislation does not define justifiable economic and social development. This, however, is alluded to in Chapter 5 from NEMA (see above) which suggests that development should 'maximise benefits'. This could be interpreted to mean maximizing social and economic benefits.

Legislation, therefore, does not provide guidance on whether development can, or cannot, be deemed to be justifiable. Without this, it is difficult to promote 'justifiable' development such as schools, housing and health facilities that ensure that rights in the South African Constitution are fulfilled, in preference to other development which does not.

Chan and Yung argue that greater flexibility should be used in the way building legislation is applied to ensure that there is greater support for land uses that are beneficial to communities or which can be 'justified' in this way. In their review of building-related legislation for Hong Kong they show that legislation tends to hamper innovation and can produce unsustainable infrastructure (Chan and Yung, 2003).

7.12 Conclusions and recommendations

This paper suggests that the South African Constitution provides an appropriate overarching legal framework within which sustainable development can be addressed in South Africa. It suggests, however, that building-related legislation does not adequately uphold and support Section 24 of the South African Constitution or ensure that sustainable development is addressed.

In particular, it argues that the rights and obligations outlined in Section 24 have not been sufficiently translated into explicit built environment requirements, or sufficiently recognized and supported, through enabling legislation. A number of recommendations are outlined below which could be explored to ensure that building-related legislation is more effective in meeting constitutional obligations.

- **Health and well being:** Minimum performance requirements such as day lighting and thermal conditions could be prescribed for buildings to ensure health and well-being of occupants was achieved. For example, prescribing minimum R-values (insulation) in roofs of buildings, which is a relatively low-cost measure, could be used to improve health and well-being of occupants in many South African homes (Mathews *et al.*, 1995).
- **Indigenous and sustainable construction:** Indigenous and more sustainable alternative construction methods should be recognized and supported through deemed-to-satisfy clauses in the building regulations. This would enable buildings with this type of construction to readily achieve approval from local authorities without additional expense.[18]
- **Sustainable development:** Sustainable development should be translated into mandatory requirements for the built environment. This could include mandatory energy and water efficiency measures as well as prescribed labour intensity, health and education performance requirements.[19]
- **Justifiable development:** Justifiable development could be defined in legislation and steps taken to promote this rather than development that was less justifiable or unjustifiable. This legislation could be used to ensure that development that fulfils rights defined in the constitution, such as the right to housing, education and health is prioritized.

While these recommendations could improve building standards, the law is still likely to be seen as a rule-based system dictated by government. Building-related legislation will continue to define only minimum standards and, in turn, may be seen as largely irrelevant to the design process (Patlis, 2005).

In order for sustainability and constitutional requirements to be more effectively integrated into buildings, a more explicit and practical understanding of their implications for the built environment needs to be developed. This should be used to inform decision-making processes in order to achieve built environments that are more sustainable and reflect the requirements of the South African Constitution more fully. For instance, instead of merely stating maximum energy consumption requirements, legislation could be used to show the implications of different design decisions in order to encourage the selection of innovative and appropriate solutions that not only enable energy efficiency targets to be achieved but also ensure that other sustainable development priorities such as employment are addressed. The law from this perspective would enable and support a process of integrated decision-making guided by practical norms based on human rights and sustainability (Reisman and Aaron, 1987).

Notes

1 Section 9 of the South African Constitution.
2 Section 36 of the South African Constitution.
3 Section 7 of the South African Constitution.
4 Section 24 of the South African Constitution.
5 Principle 1 of the Declaration of the United Nations Conference on the Human Environment.
6 Article 24 of the African Charter on Human and People's Rights.
7 Preamble to the Constitution of the World Health Organization.
8 Principle 3 and 4 of the Rio Declaration on the Environment and Development.
9 Section 26 of the South African Constitution.
10 Section 27 of the South African Constitution.
11 Section 29 of the South African Constitution.
12 The SABS stands for the South African Bureau of Standards, a standards generating body.
13 The CSIR stands for the Council for Scientific and Industrial Research, a research organization.
14 Agrément Certificates are issued by Agrément South Africa, a building materials certification agency.
15 Chapter 1 of the National Environmental Management Act.
16 Chapter 5 of the National Environmental Management Act.
17 Chapter 5 of the National Environmental Management Act.
18 The Code of Practice for the Construction of Dwelling Housings in Accordance with the National Building Regulations developed by the South African Bureau of Standards in 1989 could be used as basis for amending regulations.
19 The SANS 204 Energy Efficiency in Buildings standard, developed by South African Bureau of Standards in 2008 could be used as a basis for energy efficiency legislation for buildings.

References

American Society of Heating, Refrigerating and Air-Conditioning Engineers (1992) *ANSI/ASHRAE Standard 55-92 Thermal Environmental Conditions for Human Occupancy*, Atlanta, GA.

Chan, E. W. and Yung, E. K. (2003) "Is the development control legal framework conducive to a sustainable dense urban development in Hong Kong?", *Habitat International*, Vol. 28, Issue 3, pp. 409–426.

Department of Environment and Tourism (1998) *National Environmental Management Act*, Department of Environment and Tourism, Pretoria.

Department of Environment and Tourism (2009) *The National Climate Change Response Policy*. Department of Environment and Tourism, Pretoria, p. 8.

Department of Labour (2003) *Occupation Health and Safety Act, Construction Regulations*, Department of Labour, Pretoria.

Department of Labour (1987) *Environmental Regulation for Workplaces*, Department of Labour, Pretoria.

Department of Labour (1988) *Facilities Regulations*, Department of Labour, Pretoria.

De Waal, J., Erasmus, G. and Currie, I. (1999) *The Bill of Rights Handbook*. Juta, Johannesburg, p. 393.

Dolley, C. and Hermanus D. (2009) 'Fire rages as Cape boils'. *Pretoria News*. 6 March, p. 1.

Mathews, E. H., Richards, P. G, van Wyk, S. L. and Rousseau P. G. (1995) 'Energy Efficiency of Ultra-low Cost Housing', *Building and Environment*, Volume 30, Number 330, pp. 427–442.

Patlis, J. (2005) 'The role of law and legal institutions in determining the sustainability of integrated coastal management projects in Indonesia', *Ocean & Coastal Management*, Vol. 48 Issues 3–6, pp. 450–467.

Reisman, M. W. and Aaron, M. S. (1987) *Jurisprudence: Understanding and Shaping the Law*, Yale University Press, New Haven.

South African Bureau of Standards (1990) *Code of Practice for the Application of the National Building Regulation*, South African Bureau of Standards, Pretoria, pp. 101–115.

South African Bureau of Standards (1990a) *Code of Practice for the Application of the National Building Regulation*, South African Bureau of Standards, Pretoria, pp. 101–115.

South African Bureau of Standards (1990b) *Code of Practice for the Application of the National Building Regulation*, South African Bureau of Standards, Pretoria, p. 40.

South African Bureau of Standards (1990c) *Code of Practice for the Application of the National Building Regulation*, South African Bureau of Standards, Pretoria, p. 43.

South African Constitutional Court (2009) 'What is the definition of a constitution?' available at: http://www.constitutionalcourt.org.za/site/home.htm (accessed June 2009).

South African Human Rights Commission (1996a) *Reflections on Democracy and Human Rights*, South African Human Rights Commission, Johannesburg, p. 119.

South African Human Rights Commission (1996b) *Reflections on Democracy and Human Rights*, South African Human Rights Commission, Johannesburg, p. 120.

South African Human Rights (1996c) *Reflections on Democracy and Human Rights*, South African Human Rights Commission, Johannesburg, p. 120.

Sustainable Energy Africa (2006) *State of Energy in South African Cities*, Sustainable Energy Africa, Cape Town.

Terblanche, P. (1992) *The Health and Pollution Dimensions of Domestic Energy Sources, Atmospheric Impact Management*, CSIR Report, Pretoria.

Wieslander G., Norbäck D., Björnsson, E., Janson, C. and Boman, G. (1996) 'Asthma and the indoor environment: the significance of emission of formaldehyde and volatile organic compounds from newly painted indoor surfaces'. *International Archives of Occupational and Environmental Health*, Volume 69, Number 2, pp. 115–124.

8 Energy efficiency in buildings and building control regulations in South Africa

Joachim E. Wafula, Kennedy O. Aduda and Alfred A. Talukhaba

8.1 Introduction

Increasing energy demand has been identified as one of the key contributors to global warming (UNEP, 2003). Energy efficiency in buildings is emerging as one of the mitigating factors of climate change. In the context of buildings, energy efficiency is defined as the ability to provide the required internal environment and services with minimum energy use in a cost-effective and environmental friendly manner (CIBSE, 2004). While defining energy efficiency, it is imperative that it must be distinguished from energy intensity. It must be remembered that energy intensity is the ratio of energy consumption to some measure of demand for energy services and remains preferable as an indicator for energy efficiency (EIA, 2000). The key attributes considered as a prerequisite to energy efficiency are reduction in energy output, maximization of work output and overall cost reduction.

For South Africa, energy efficiency has become highly critical considering that projected electricity demand nearly equals the current production levels. Eskom (2008) reports that South Africa's electricity demand is forecasted to be 48,624 MW in the year 2014 against an operating capacity of 50,324 MW. This highlights the key issue of overall access to clean energy as it can be clearly noted that the existing reserve is currently very low. In their annual report for the business year 2005, the main electricity producing company in South Africa, Eskom, openly declared that the country's electricity production reserve was below the world's recommended level which is 15 per cent (Eskom, 2007). Figure 8.1 illustrates the issue of electricity demand and capacities in South Africa; it can be observed that the electricity production reserve has been dwindling in the last ten years and that at the moment the peak demand is almost at par with the operational capacity of the generation plants.

The end result is unscheduled load shedding as well as planned load shedding. This underscores the energy poverty issue in South Africa, where there is significant expansion in demand and supply coupled with a lack of reliable access. In order to effectively solve this problem South Africa must

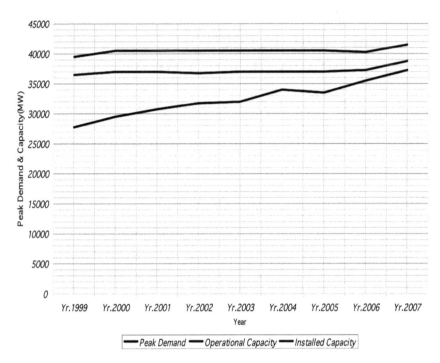

Figure 8.1 South Africa's electricity demand and capacity

Source: Adapted from South Africa's Government, National Resonse to S.A. Electricity Shortage.

improve on energy efficiency, introduce demand scheduling and undertake to use alternative energies as well as strategically increase its supply of energy.

All these methods are significant though the strategic demand reduction, especially via the practice of energy efficiency, seems the most effective and cost-efficient method. This is because it involves the use of the same amount of electricity input to undertake greater amount of tasks/work. It may encompass the use of design, operation and management principles that are more efficient in energy utilization. As a result the idea of green building and energy conservation are set to feature prominently in the path to increase reliable energy access.

It should be noted that electricity demand reduction is made difficult by the perceived low price of power, lack of knowledge and understanding of the concepts, institutional barriers and resistance to change as well as lack of investors' confidence due to the lengthy payback period which compounds the problem. The South African Energy Efficiency Strategy remains focused on the overall electricity demand reduction of 12 per cent by the year 2015 as a means of alleviating energy poverty (Department of Minerals and Energy

(DME), 2005). The promise of increased access to electricity through demand reduction makes a convincing argument for energy efficiency studies.

While looking at capacities and demand for energy one should not lose focus of the fact that South Africa's energy sector is reputed for its high carbon dioxide intensity. This is evident in recent reports on global warming outlined in the 2007/2008 United Nations Development Programme (UNDP) human development report entitled 'Energy development and climate change; decarbonising growth in South Africa', which observed that by international standards, South Africa's economy is extremely high energy intensive in terms of energy consumption in relation to its Gross National Product. The UNDP's report (2007) further stated that despite its relatively low GDP (52nd in the world) and low human development index (121st in the world), South Africa had relatively high green house gas emissions (37th in overall carbon dioxide gas emissions according to the 1999 data supplied by the department of tourism and environment).

This is mainly due to the fact that coal remains the dominant energy supply source at 70 per cent of the country's primary energy followed by electricity and natural gas at 28 per cent and 2 per cent respectively; in addition, coal fuels 93 per cent of electricity production (DME, 2005). Available data indicate that South Africa's energy sector remains the highest contributor to carbon dioxide gas emissions to the atmosphere at approximately 70 per cent (UNDP, 2007). This implies that any demand reduction as a result of energy efficiency would greatly contribute to a reduction in green house gas emissions, especially carbon dioxide (CO_2).

8.2 Energy efficiency in buildings

UNEP (2007) observes that buildings account for between 30 and 40 per cent of all primary energy worldwide. In South Africa, the building industry accounts for 27 per cent of electricity use and 12 per cent of the final energy use (DME *et al.*, 2002). Figures 8.2 and 8.3 illustrate details of energy use in South Africa.

Building energy can be categorized according to the phase of energy use. Jones (1998) and later Sartori and Hestness (2007) explain that building energy can be of embodied, grey, induced, operation or demolition-recycling type (Jones, 1998; Sartori and Hestnes, 2007). In a study done for office buildings in the United States, Japan and Finland, Jumilla (2004) summarizes various energy phases for buildings in Figure 8.4.

It can be seen that the bulk of energy use in buildings is operation energy (Jumilla, 2004). As such it is the energy that makes the building habitable and includes consumptions on heating, ventilation, lighting and air conditioning (Sartori and Hestnes, 2007; Jones, 1998). Comparatively, demolition/recycling energy and construction energy contribute negligibly to the final energy use in the life cycle of a building (refer to Figure 8.4). As such, energy efficiency becomes an automatic choice for energy management in buildings.

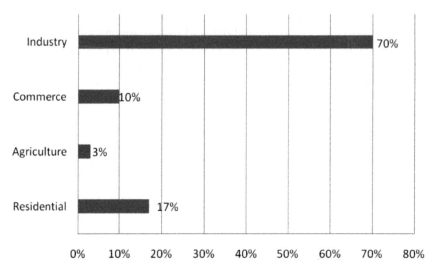

Figure 8.2 Electricity demand by sector in SA
Source: Adapted from DME *et al.*, 2002.

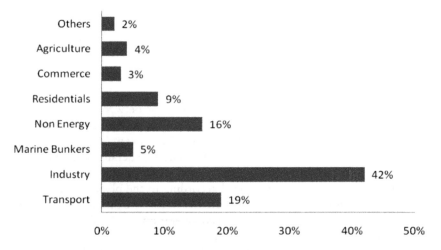

Figure 8.3 Final energy demand by sector in SA
Source: Adapted from DME *et al.*, 2002.

The main drivers for energy efficiency are the twin issues of economic returns due to savings in energy use and reduction in the emissions of CO_2 gas in the atmosphere as explained earlier. In South Africa, the Energy Efficiency Strategy paper asserts that a 25 per cent saving is possible in the commercial and public building sector (DME, 2005). Liang *et al.* (2007) report that

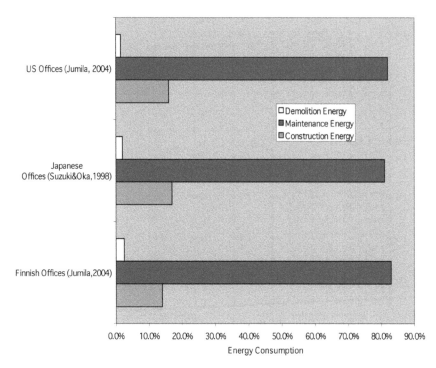

Figure 8.4 Energy use by life cycle for selected cases
Source: Adapted from Jumilla, 2004.

improving energy efficiency in buildings is one of the most cost-effective ways of reducing greenhouse gas emissions, particularly CO_2. However, it must be remembered that energy efficiency programmes in buildings must be properly focused on both new and existing buildings alike.

8.3 Building regulations and control in South Africa

Several instruments or options are available towards the promotion of energy efficiency in buildings. Van Egmond (2001), OECD (2003), Kuijsters (2004) and Reinink (2007) separately affirm these options as regulatory, fiscal, economic, or communications instruments; these instruments may take the form of regulatory instruments, indirect regulation or information instruments (see Figure 8.5). It is worth noting that the choice of these instruments by governments is determined by the level of success that is achievable when applied. This is, however, variant to the socio-economic and political context of the locality. It is noted, as an example, that implementation of the building codes have reduced energy consumption of new dwellings in the USA by about 30 per cent. The Chinese government have

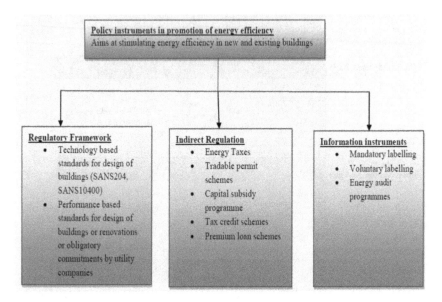

Figure 8.5 Instruments for energy efficiency implementation in buildings
Source: Adapted from OECD, 2003.

established an energy consumption target in buildings which is 65 per cent less than the current practice in existing buildings, via an 'energy consumption standard' for the construction sector whose compliance is encouraged by tax and fees rebate systems for low-energy buildings to encourage their construction (UNEP, 2007). On the other hand, energy regulations in buildings in the Netherlands reduced consumption rates by 15 per cent on introduction in 1995 and later by 27 per cent on tightening of requirements (Ecofys, 2004). Other countries where the introduction of energy regulations have been successful are Denmark (Act to Promote Energy and Water Savings in Buildings)[1]; and India where the Energy Conservation Act was enacted in 2001.[2]

Regulation governing energy efficiency in buildings in South Africa is still in infancy. It is currently to be found primarily under the Electricity Regulation Act,[3] which stipulates that the regulator shall take into account the energy efficiency measures undertaken by the client while deciding on tariff structure and the electricity regulations for compulsory norms and standards for reticulation services.[4] It must be noted, however, that building controls and regulations with regards to energy efficiency may be through voluntary implementation or mandatory implementation or may entail a mixed approach (see Figure 8.5).

The regulatory framework for buildings in South Africa is defined by the National Building Regulations and Building Standards Act 103 of 1977.

Section 4 of this Act and the subsequent amendments clearly vest the power of approval of buildings with the local authorities. The approval process revolves around administrative matters, such as protecting property, public health and safety and convenience for users and occupiers of the building. In addition, a daily fine of ZAR100 is prescribed by the Act as payable for any day that an illegal structure remains erected. Further, section 15 of the Act authorizes the building control officers from the local authorities to enter facilities and make assessments in relation to applications for approval.

Most importantly, the Minister is empowered to make additional regulatory sections to this Act in consultation with the local authorities. This could include the following, among others:

1 Preparation, submission and approvals of plans and specifications or alterations
2 Inspections and tests on all buildings
3 Nature of building sites
4 Provision of ventilation, artificial lighting, heating and other building services

It can be argued that the above provides an entry point for government regulations with regard to energy efficiency issues in buildings. Despite this, most local authorities in South Africa remain reluctant to incorporate energy efficiency issues in building regulations and control activities. Other requirements of the National Building Regulations include adhering to the requirements of all the prescriptive regulations and satisfying all functional regulations.

Indirect regulation is not overtly noticeable in South Africa except via the programmes of Demand Side Management (DSM) run by the electricity supply commission, Eskom. In these programmes subsidies are provided for installation of energy efficient motors and solar geysers.

8.4 Promotion of energy efficiency in buildings in South Africa

The initial phase of energy efficiency programmes by the South African government was the publishing of two white papers which culminated in the energy efficiency strategy (DME, 2004). In addition, in response to the worrying electricity supply problem, Eskom introduced a DSM measure which aimed at load reduction (Harris and Krueger, 2005). Harris and Krueger (2005) further observe that the DSM was accompanied by the imposition of electricity tariffs to bolster cash reserves for the electricity generating company and in the process provide capital for investments in future power plants.

Winkler *et al.* (2006), Winkler (2007) and Talukhaba and Aduda (2008) have observed that policies focusing on greater efficiency lead not only to energy savings, but also to cost savings. All policy interventions have upfront costs, but energy efficiency saves money over the life of the intervention. Such savings do not only have an economic benefit, they make a contribution to social development by benefiting poor households. Bennett (2001) outlines five reasons for resistance to the implementation of energy efficiency opportunities in South Africa as related to *attitude* (I know my business best), *resistance to change* (everything is going along just fine), *low cost of energy* (many users see energy as a minor input cost, relative to raw material and labour, and tend to concentrate on these), *lack of capital* and *uncertainty regarding the future.*

In the building sector, energy efficiency programmes additionally concentrated on the development of a star rating system in the proposed energy efficiency standards. This standard prescribed the energy efficiency levels per square metres of the floor area for specific tasks and different climatic areas (Harris and Krueger, 2005; Standards South Africa, 2008).

With regard to the energy efficiency savings projections in the building sector, up to 28 per cent in heating, ventilation and air conditioning (HVAC) systems and up to 100 per cent saving in lighting is expected to be realized in commercial buildings whereas 4.2 per cent saving in heating is expected in residential buildings by the year 2013 (DME, 2003). In terms of regulation, it is noted that the draft standard for energy efficiency in buildings was unveiled in 2008 and if adopted as the national standard would contribute greatly to energy efficiency in buildings; these are South Africa National Standards (SANS) 204[5] Parts 1, 2 and 3, Edition 1. The newly published building standard is expected to eventually be applied to insulation levels, solar water heaters and energy-efficient lighting and will be prescriptive; at the initial phase it will apply only to new buildings (Standards South Africa, 2008). At the same time, by 2007, the South Africa Bureau of Standards (SABS) 0400 (the Building Code) was in the process of being rewritten to be SANS 10400 to take into account the energy efficiency standards for retrofits and alterations (Reynolds, 2007; du Toit, 2007).

SANS 204 focuses on new buildings; it specifies the general requirements for design and operation of energy-efficient buildings with both natural and artificial environmental control and subsystems. The key issues of SANS 204 are as follows (Standards South Africa, 2008):

i   Maximum energy demand and maximum annual consumption are mentioned in accordance with the thirty-one classifications of occupancies of buildings and prevailing climatic conditions in the standard.

ii The standard state for building envelope design is outlined.

iii Energy rating for equipment and appliances fitted in new buildings is required to comply with requirements of SANS 10400-O

iv Purpose-driven planned maintenance of the mechanical/electrical component is advised to be in line with broader economic and energy efficiency agenda.

It is envisaged that the implementation of SANS 204 will fall under the docket of local authorities. It is noted, however, that it will take considerable time for South Africa to start realizing the fruits of building regulations. Holden (2004) approximates this to a minimum of five years.

At the moment it is noted that local authorities are ill equipped in implementing and enacting local building regulations due to legal and technical incapacity (du Toit 2007). It is envisaged that these problems will be solved by rewriting their bylaws to incorporate energy efficiency regulations such as limiting the quantity of energy consumption in buildings through setting maximum energy/m² caps and making the same regulations part of the building applications and approval procedure. In the absence of local regulatory powers, cities can consider developing local guidelines or standards. For uniformity and acceptability, these should ideally be based on existing standards and norms and best practices from similar areas.

Some cities have explored this possibility by developing local energy efficiency building regulations, but have since abandoned this process given that building regulations are established nationally in order to promote uniformity within a vast sector. However, the City of Potchefstroom has adopted an interesting local initiative whereby they have made the South African Energy and Demand Efficiency Standard (SAEDES[6]) guidelines mandatory for all new municipal buildings. It was unveiled in the 1990s as a guideline by the Department of Minerals and Energy. In the words of van der Merwe and Grobler (2003) the SAEDES document was meant to provide technical guidelines for a framework of technical performance provisions and environmental acceptability for commercial buildings.

DME (1998) and van der Merwe and Grobler (2003) observe that SAEDES provides guidance for the energy-efficient maintenance and operation of commercial buildings as well as information about the education and training of all personnel in energy efficiency. In a nutshell it is observed that the document is comprehensive, flexible, easy to use and practicable. Where commercial building applications must go through an environmental impact assessment process, the local authorities have had to introduce the SAEDES as a condition for the development.

8.5 Proposals on implementation of energy efficiency in buildings in South Africa

In a departure from the established norm, the Building Research Establishment (2008) proposes that compliance with mandatory minimum energy performance requirements for buildings should be confirmed by formally certified private assessors and paid for by the building owners. It is further suggested that the process be audited by the authority under which the code is issued (Building Research Establishment, 2008). Thus the following enforcement model is idealized by RICS (see Figure 8.6). In South Africa, however, the local authorities have a well established buildings control and regulations framework and structure and there is a deliberate intention by government to retain and create employment at local levels. Therefore it is proposed that the enforcement and the certification be done by these local authorities, who are mandated to enforce building regulations within their areas of jurisdiction as per the Act (103 of 1977) (see Figure 8.7).

In 2009 the Department of Energy succeeded in pushing through two regulatory frameworks which could also have an impact on energy efficiency regulations in buildings. There was an amendment of the Electricity Regulation Act (4/2006-Schedule 2) that stipulates that the regulator shall take into account the energy efficiency measures undertaken by the client while deciding on tariff structure (Republic of South Africa, 2008). In addition, a further amendment of the same Act entitled 'electricity regulations for compulsory norms and standards for reticulation services' makes it mandatory for end users with monthly consumption above 1000 kWh to have smart metering devices to benefit from the time of use tariff by 1 January 2012 (Republic of South Africa, 2009).

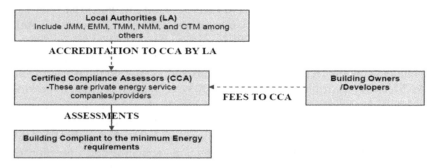

Figure 8.6 RICS compliance enforcement model

Source: Adapted from Building Research Establishment, 2008.

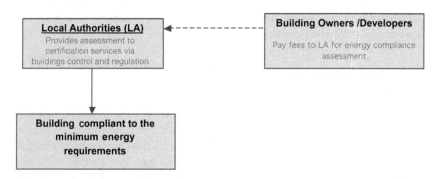

Figure 8.7 Proposed compliance enforcement for the local authorities in South Africa

Source: Authors' construction

8.6 Worldwide scenarios

Around the world, building controls and regulations with regard to energy efficiency are evolving and are being implemented via both voluntary and mandatory standards or may entail a mixed approach. Generally, in 2000 it was reported that the majority of countries in South America and Africa did not have energy efficiency regulations for buildings (Janda and Busch, 1994). During this period nearly all the central European countries had voluntary energy efficiency regulations in buildings except for a few countries which had mandatory energy efficiency regulations (Janda and Busch, 1994). Specifically, Scandinavia uses a general norm of applying national building codes and standards, which regulate physical, thermal and electrical requirements of building components, service systems, indoor conditions, health and safety standards, operation and maintenance procedures and energy calculation methods (UNEP 2007). A number of building codes currently include energy performance standards, limiting the amount of energy that buildings can consume. The building codes in Scandinavia are enforced by the various building development control agencies in the respective countries.

It is reported that currently the renewable energy and energy efficiency partnership (REEP) project aims to promote the concept of low-energy buildings in China. The results will be incorporated in proposed new legislation regarding low-energy buildings by the Chinese government in order to ensure that appropriate policies and building codes are implemented to encourage and deliver the required reduction in energy consumption (UNEP, 2007). Indeed, implementation of elements of these codes are to be piloted by the four major cities of Beijing, Shanghai, Tianjin and Chongqing and other economically developed big cities such as Shenzhen (Liang *et al.*, 2007).

The European Union often uses directives for regulating various environmental themes whose implementing agents are various government departments in the respective member states. In most cases this falls in the ambit of local authorities in the respective countries. The directive on the energy performance of buildings was enacted in January 2003 and its main elements are specified as follows (UNEP 2007):

i   Minimum energy performance requirements, for new buildings and for major renovation of existing buildings larger than 1000m^2;
ii Energy performance certificates to be made available when buildings are constructed sold or rented out;
iii The year 2010 is the reference year after which the rules will be extended to apply to all buildings and renovations(at the moment buildings below 1000 m^2 are not covered).

8.7 Conclusions and recommendations

From the foregoing literature it is evident that several developing countries have already enacted legislation on energy efficiency in buildings. However, only a few evaluations or studies are available to show the best way to apply this legislation to achieve energy efficiency goals in buildings. Due to late entry into energy efficiency practice, most developing countries lack quantitative data and are mainly reliant on data from the developed nations; this poses a contextual problem in that their programmes may not be applicable to the prevailing local condition. This can be seen in the case for South Africa which borrowed quite heavily from the Nordic countries in formulating its energy efficiency codes in buildings (Reynolds, 2007).

Developed countries, however, find solace in the fact that increasing energy prices will continue to be the catalytic drive for improved energy efficiency policies in developing countries. In South Africa the main electricity utility company Eskom successfully lobbied to raise tariffs by nearly 32 per cent in 2008 and later succeeded in getting permission to further increase the tariffs in 2010 (Engineering News, 2008; South Africa Online, 2010). Indeed, in February 2010, the National Energy Regulator of South Africa approved Eskom's application to increase tariffs by a nominal 24.8 per cent for the 2010/2011 year, 25.8 per cent for the 2011/2012 year and 25.9 per cent for the 2012/2013 year (South Africa Online, 2010). This would eventually lead to an average electricity price increase from a current R0.33/kwh to R0.415/kwh this year, then go up to R0.52/kwh in 2011/2012 and finally to R0.65/kwh in 2012/2013, thus doubling the average rate in the next three years.

As a result it is expected that the South African government will, in response, empower the local authorities as the electricity distributing agencies to incorporate energy efficiency measures in their bylaws apart from engaging in advocacy and training of the built environment practitioners and general citizenry. It should, however, be emphasized that the success of the Building Regulations is highly dependent on compliance by the construction industry hence the need for local authorities to invest in continuous training and advocacy of building control officers and built environment professionals on the energy efficiency regulations.

It is also expected that energy security considerations and rapidly rising energy demand like that which contributed to an unstable electricity supply in South Africa during the years 2006 to 2008 will prompt the local government and utilities companies to act in promoting energy efficiency. This is already evidenced in the enactment and amendments to the two electricity Acts in South Africa – the Electricity Regulation Act and the Electricity Regulations for Compulsory Norms and Standards for Reticulation Services – which were mentioned earlier.

In terms of mode of policy implementation a leaf must be borrowed from developing countries such as Malaysia, Brazil, Morocco and partly Thailand which first introduced voluntary standards for buildings then progressively made them mandatory (UNEP, 2007). It should be noted, however, that although best practices and experiences can be shared and regional co-operation is useful, building code specifications cannot be uniform for all parts of a country due to climatic and other peculiar differences. This further fosters the idea of local authorities as the ideal enactor and enforcer for energy efficiency codes in buildings.

This chapter has demonstrated the effectiveness and success of regulatory and control mechanisms like energy efficiency building codes when enforced by the local authorities. It is therefore not only natural but a matter of necessity for South Africa to use the local authorities as the enforcing agents for energy efficiency codes in building. This is made easier by the fact that in South Africa the local authorities are considered the most important party in service delivery (Republic of South Africa, 1995). It is noted that in South Africa, energy efficiency policies are in various government agencies and departments (these are the Department of Minerals and Energy, Eskom, the National Energy Regulator (NERSA) and the Department of Public Works, Roads and Transport). Policy implementation is therefore a challenge. For the building sector, it is recommended that the local authorities should take the initiative and learn from the city of Potchefstroom and effect energy efficiency regulations through their building control and approvals regulations processes and systems.

While implementing the energy efficiency building regulations through their building regulations and control mechanisms, local authorities should make appropriate education and outreach programmes to raise awareness of the general public regarding energy efficient buildings, and demonstrating options for retrofitting of existing buildings as well as the design and construction of new buildings.

Existing building stock in particular provides a huge opportunity for energy savings. Most existing buildings were not designed for energy efficiency, but by retrofitting with up-to-date products, technologies and systems, a typical building can realize significant energy savings. In South Africa, the end of Apartheid also signalled an end of an era of cheap energy. This, coupled with the need to create more construction work in the building industry following the end of World Cup–associated works, makes it natural that retrofits are going to be an important driver of the construction work in the economy.

For example, local authorities can get direct and assured energy efficiency benefits in retrofits by enacting and implementing high-yield specific design regulations which require no additional cost. This may include a planning tool like the addition of vegetation requirement in an already built up area. This will lower temperatures during summer and reduce heating costs. As new technologies develop, many other specific cost-effective design elements will present opportunities for targeted mandates in retrofits. Changing from conventional hot water heating systems to solar water heaters in domestic buildings is another key regulation which should be implemented urgently in retrofits. Heating water can take up to 15–25 per cent of the energy use in a home and a solar water heater can cut annual operating costs by up to 50–80 per cent. In 2000, Barcelona implemented such a requirement and was subsequently followed by other cities all over Spain successfully.

For the case of the existing stock owned by the municipalities and the local authorities themselves, operational practically requires that they examine them first for energy efficiency and retrofit them appropriately as pilot/case projects. There are many obvious resources which can be explored in detail to deal with this. One stand-out practice is the employment of the Energy Service Company (ESCO). The ESCO develop, design and finance the energy efficiency project/program on behalf of the local authority on its premises. They install and maintain the energy efficiency equipment involved, measure, and monitor and verify the programs energy savings. They also assume the financial risk that the program will deliver the amount of energy savings guaranteed. This is appropriate for the local authorities since they do not have upfront costs and the ESCOs get paid from the savings generated by the program. This allows the local authorities to deliver their normal operations without additional financial burden for energy efficiency programs on their building stock. Significantly, results of these programs can subsequently be rolled out to all other large building stock holders in their jurisdiction. The challenge to the local authorities in this arrangement lies in establishing appropriate procurement legislation which protects its interests at all times. For South African local authorities, this is an attractive and fiscally sound means of financing an energy efficiency upgrade/retrofit in local authority buildings.

Energy efficient retrofits/renewals and refurbishments do involve a high financial outlay. Getting appropriate finance for it is therefore important to its success. In South Africa, these interventions in refurbishments could be viewed as reducing demand. This should then be eligible for funding from sources like the Eskom DSM Fund. Appropriate partnerships should be created between the local authorities' building regulations control entities and the above funds administration systems to make it mandatory for the recommendations of the local authorities to be part of the criterion used for developers to draw from it. However, the South African local authorities can develop additional revenue streams to support energy efficiency new and renewal/retrofit programs in their jurisdictions and compliment the Eskom DSM programs. In a scheme that can serve both to raise funds and promote installation of renewable energy, the local authorities can introduce a fee, of say R5,000 for all new developments and large retrofits of over 2,000 square feet if they fail to include the installation of a 2kW solar photovoltaic system or equivalent renewable energy system. Similarly, homeowners who consume energy beyond reasonably forecasted budgets, especially on energy intensive activities such as heating outdoor pools or spas can be charged a mitigation fee of say R5,000 if they fail to install energy efficiency or renewable energy systems. The funds collected can be used to promote energy efficiency and renewable energy policies in buildings. Also, the South African policy makers should look at the Property Assessed Clean Energy (PACE) financing mechanism as a tool in overcoming barriers to commercialized building retrofits (Pike Research, 2010). This way, the goal of energy efficient retrofits will have made a giant step forward.

It is acknowledged that research gaps exist in several areas of energy efficiency in buildings in South Africa; an example is the lack of costs estimation data for energy-efficient buildings. Similarly, there is no efficient market guidance for energy-efficient buildings. This causes low sensitivity to energy-efficient buildings issues in the design, construction and maintenance of buildings. Research and studies in these areas will enhance learning and help make improvements in programme designs for energy-efficient buildings in the future.

8.8 Acknowledgement

The authors acknowledge financial support for the study from the South African National Energy Research Institute (SANERI).

Notes

1 Danish Energy Authority (DEA) 1996 Act to promote energy and water savings in buildings, no. 485 of 12 June 1996, legislative document.
2 Government of India, 2001, Energy Conservation Act, 2001, September 2001.
3 Republic of South Africa, 2009, Electricity Regulation Act, 2006 Notice 139 of 2009, Government Printers, Pretoria.
4 Republic of South Africa, 2008, Electricity Regulations for Compulsory Norms and Standards for Reticulation Services, Government Printers, Pretoria.
5 Standards South Africa (2008) SANS 204-1:2008 Edition 1, Pretoria, South Africa.
6 Department of Energy and Minerals (DME) 1998, SAEDES guideline: Energy and demand efficiency standard for existing and new commercial buildings, Pretoria.
7 Department of Energy and Minerals (DME) 1998, SAEDES guideline: Energy and demand efficiency standard for existing and new commercial buildings, Pretoria.

References

Bennett, K. F. 2001. Energy efficiency in Africa for sustainable development: A South African perspective. UNESCO Workshop on Sustainable Development, Nairobi.
Building Research Establishment 2008. Can building codes deliver energy efficiency? Defining a best practice approach, A report for the Royal Institution of Chartered Surveyors. Available at http://www.rics.org/site/download_feed.aspx?fileID=4128 &fileExtension=PDF.
Chartered Institution of Building Services Engineers London (CIBSE) 2004. Energy

Efficiency in buildings CIBSE Guide F. Second Edition. Page Bros. (Norwich) Ltd, Norfolk, pp. 18.8, 19.6–19.7.

Department of Energy and Minerals (DME) 1998. SAEDES guideline: Energy and demand efficiency standard for existing and new commercial buildings, Pretoria.

Department of Minerals and Energy (DME), Eskom, and Energy Research Institute-University of Cape Town 2002. Energy Outlook for South Africa. Available at http://www.dme.gov.za/pdfs/energy/planning, pp. 1–41.

Department of Minerals and Energy (DME) 2003. Free Basic Alternative Energy Policy: Households Energy Support Programme.

Department of Mineral and Energy (DME) 2004. Draft Energy Efficiency Strategy of the Republic of South Africa.

Department of Minerals and Energy (DME) 2005. Energy efficiency Strategy of the Republic of South Africa. Available at http:/www.dme.za.

Du Toit, E., 2007. How to Implement Renewable Energy and Energy Efficiency Options. Sustainable energy Africa, Westlake.

Ecofys, 2004. Evaluatie van het klimaatbeleid in de gebouwde omgeving 1995–2002, prepared for department of housing, spatial planning and environment. Available from: www.ecofys.nl/nl/publicaties/documents/samenvattingrapport 20040623.pdf (accessed 22 April 2009).

Energy Information Administration (EIA) 2000. Energy efficiency and measurement. INTERNET.http:/www.eia.doe.gov. Accessed on 15 February 2008.

Engineering News, 2008. Nersa publishes municipal electricity tariff increase details, 8th July 2008. Available at http://www.engineeringnews.co.za/article. Accessed on 31 July 2009.

Eskom, 2007. Partnering growth-Director's Annual Report 2007. Available at http://www.eskom.co.za/annreport.07. Assessed on 20 February 2008.

Eskom, 2008. Annual report 2008: Together rising to the challenge. Available at http://www.eskom.co.za/annreport08/info_sheets/005.htm. Accessed 31 July 2009.

Harris H. C. and Krueger D. L. W., 2005. Implementing energy efficiency policy in housing in South Africa, *Journal of Energy in Southern Africa*, Vol. 16, No. 3, August 2005.

Holden, R. M., 2004. Towards Energy Efficient Building Regulations – The City of Johannesburg's Building Bylaws Review Process, City of Johannesburg, Johannesburg.

Janda, K. B. and J. F. Busch, 1994. Worldwide Status of Energy Standards for Buildings, *Energy Volume* 19, pp. 27–44.

Jones D. L., 1998. Architecture and the Environment. London, Laurence King Publishing.

Jumilla S., 2004. An Environmental Impact of an Office Building throughout its Life Cycle. Espo, Helsinki University of Technology, Construction Economics and Management (Doctoral Dissertation) Quoted from United Nations Environmental Programme (2007) Buildings and Climate Change: Status, Challenges and Opportunities. Available at http://www.unep.fr/pc/sbc/docu. p. 8.

Kuijsters, A., 2004. 'Environmental response of the Chilean building sector: Efforts and Constraints towards environmental building practices in the Santiago Metropolitan Region', MSc Thesis, Eindhoven University of Technology. Available from: http://www.tue.nl/bib on 2009/05/08.

Liang Jing, Li Baizhan, Wu Yong, and Yao Runming, 2007. An investigation of the existing situation and trends in building energy efficiency management in China. Energy and buildings. Elsevier Science, Beijing.

Organisation for Economic Co-operation and Development (OECD), 2003. Environmentally sustainable buildings: Challenges and policies, Paris, OECD. Available from: http://www.oecd.org/publications/pol_brief. Accessed on 15 April 2009.

Pike Research, 2010. *Executive Summary: PACE financing for commercial buildings–property assessed clean energy financing for energy efficiency retrofits and renewable energy: market opportunity, GHG reduction, and job creation.* Available at http://www.pikeresearch.com/wordpress/wp-content/uploads/2010/06/PACE-10-Executive-Summary.pdf.

Reinink, M. W., 2007. 'Towards an effective energy labelling programme for commercial buildings: A Comparative evaluation of the green buildings for Africa programme in relation to international experience', MSc Thesis, University of the Witwatersrand.

Republic of South Africa, 1995. White paper on transformation of the public service, Notice No. 1954 OF 1994, Government Printer.

Republic of South Africa, 2009, Electricity Regulation Act, 2006 Notice 139 of 2009, Government Printers, Pretoria.

Republic of South Africa, 2008, Electricity Regulation Act, 2006. Electricity regulations for compulsory norms and standards for reticulation services, Government Printers, Pretoria.

Reynolds, L. K., 2007. The South African Energy Efficiency Standards for Buildings, at http://www.cpcut.ac.za. Accessed on 20 May 2009.

Sartori I. and Hestnes A. G., 2007. Energy use in the life cycle of conventional and low-energy buildings: A review article. *Energy and Buildings* Vol. 39, No. 3, pp 249–257.

Standards South Africa , 2008, SANS 204-1:2008 Edition 1, Pretoria, South Africa.

South Africa Online(2010) Tariff increases. Available at http://www.southafrica.co.za/2010/03/18/tariff-increases/.

Talukhaba, A. A. and Aduda, K. O., 2008, "Optimisation of HVAC Systems for Energy Efficiency in Buildings", in proceedings of the 3nd Built Environment Conference, Cape Town, South Africa, 6–8 July.

United Nations Development Programme (UNDP), 2007. Human Development Report 2007/2008-Energy Development and Climatic Change; Decarbonising Growth in South Africa. Available at http:/hdr.undp.org. pp. 6–9.

United Nations Environmental Programme(UNEP), 2003. Sustainable building and construction, facts and figures, *Industry and Environment*, Vol. 26, No. 2–3.

United Nations Environmental Programme (UNEP), 2007. Buildings and Climate Change: Status, Challenges and Opportunities. Available at http://www.unep.fr/pc/sbc/docu. pp 1–87.

van Egmond- de Wilde de Ligny, ELC, 2001. Technology policies in developing Countries. Lecture Notes course ON470, Eindhoven University of Technology, Eindhoven.

van der Merwe, C. A. and Grobler L. J., 2003, The final evaluation of the South African energy and demand efficiency standard, Domestic Use of Energy Conference.

Winkler, H. E., Borchers, M., Hughes, A. G., Visagie, E. F. and Heinrich, G. S., 2006.

168 *Joachim E. Wafula, Kennedy O. Aduda and Alfred A. Talukhaba*

Policies and scenarios for Cape Town's energy future: Options for sustainable city energy development. *Journal of Energy in Southern Africa*, Vol. 17, No. 1, pp. 28–41.

Winkler, H., 2007. Energy policies for sustainable development in South Africa, Energy Research Centre, University of Cape Town.

Part IV

9 The extended Australian urban dwelling: key issues relating to private open space in expanding residential surburbia

Andrew H. Kelly and Stuart J. Little

9.1 Introduction

The garden is a fundamental element of the residential environment, especially in low density suburbia. In heavily urbanized Australia, front and back yards play a crucial role for citizens seeking their own private open space experiences. Due to sunny attractive climes, many houses extend *into* the garden – both physically and ideologically – providing a key function in everyday life. The garden is part of the house itself. As Hall puts it (2007, 27), the back garden 'can be characterised as an outside room'. It is where family and social activities take place, such as cricket games, barbecues and lazing under shady eucalypts. Depending on the householder, this may broaden to, *inter alia*, built-in playgrounds, vegetable growing and/or planting and maintaining indigenous vegetation (Head and Muir, 2007). It is also used for utility purposes, such as drying clothes and providing water tanks. Private gardens, however, are becoming smaller with enlarging residential density (Syme *et al.*, 2001; Hall, 2007). Ball games and picnics in newer suburbs may now head for local parklands.

Although the above observations apply across all Australian urban landscapes, this chapter focuses on Sydney's periphery. Australia's biggest city is surrounded by national parks to the north and south, the Pacific Ocean to the east and generally undulating lands towards the Blue Mountains to the west. Apart from (1) the rapid vertical residential expansion in central Sydney and key suburban hubs, including transport nodes, and (2) the surge of in-fill development across all residential areas, apart from some environmentally sensitive lands, housing is marching west. Despite a growing variety of residential forms across Sydney and ongoing criticism against urban sprawl, most suburban residents are proud of their home patch (Davison, 1994).

As western Sydney continues to grow, three critical issues demand scrutiny. All relate to private residential open space:

1   enhancing and protecting amenity;
2   conserving biodiversity; and
3   minimizing threats from bushfire.

This chapter will address each one below, with attention to local government which is at the forefront of land use regulation and community involvement. This third sphere of government suffers from no formal recognition in the Australian Constitution, which primarily distributes legislative power between the Commonwealth and the states. While local government is a creature of state parliaments, it is nevertheless embedded in Australian governance.

9.2 Statutory town planning in New South Wales and the urban periphery

The first comprehensive planning legislation in NSW occurred in 1945 with insertion of Part XIIA into the then *Local Government Act 1919* (NSW), which went far beyond building and subdivision control. The regime enabled the making of statutory planning scheme ordinances (PSOs), mainly to provide zoning regulations. Curiously, although the Australian planning legislation relied heavily on the UK's system, Britain followed a different pathway soon afterwards in 1947. Australian jurisdictions remained glued to stringent zoning patterns to combat conflicting land uses. But due to inadequate resources and limited outlooks, councils were originally slow to adopt planning as a vital function. Instead, it was the state government that drove metropolitan planning through the Cumberland County Planning Scheme (CCPS) of 1951, the first and only comprehensive statutory plan for the Sydney region. The CCPS planned for post-war urban expansion while advancing the 'Australian suburban dream' (Alexander, 2000, 102): i.e. detached houses surrounded by large well-watered lawns alongside almost identical townscapes.

In 1980, Part XIIA was replaced by the *Environmental Planning and Assessment Act 1979* (NSW) (EPAA). This reflected the emergence of 'modern environmentalism', encouraging innovative plan-makers to move beyond land use conflict. In addition, local communities demanded greater input in plan preparation and implementation (Roddewig, 1978). Another factor was regional planning, acknowledging that many issues apply across administrative and arbitrary borderlines. This paralleled a cascade of strategic regional non-statutory instruments across Sydney's myriad of council areas. The City of Cities: A Plan for Sydney's Future Sydney (more commonly known as the 'Metropolitan Strategy'), was introduced in 2005 with its various existing and forthcoming sub-strategies. This was more recently revised by the Metropolitan Plan for Sydney 2036, issued in December, 2010. Adjacent to this is the Growth Centres policy,[1] wherein the State Government has promised an ongoing supply of land for low density homes for Sydney's West. This reflects a fierce political push for residential expansion at Sydney's periphery. Its implementation is currently taking place via the North-West and South-West Growth Centres, with the latter predicted to accommodate about 110,000 new homes.

In terms of statutory plans, the EPAA introduced a series of statutory 'environmental planning instruments' (EPIs). At the time of writing, these have been reduced to State Environmental Planning Policies (SEPPs) and Local Environmental Plans (LEPs). SEPPs deal with matters of state or regional significance. A relevant example is SEPP (Sydney Regional Growth Centres) 2006, which provides the statutory basis for the two residential sectors mentioned above. The LEP, however, is the fundamental EPI that local government prepares and implements. As a result of legislative and policy reforms in 2006, the state government introduced the LEP 'standard instrument', commonly known as the 'LEP template'.[2] Each of the 152 councils across NSW must abide by the template in redesigning its own LEP. Numerous 'templatized' LEPs have yet been gazetted. By providing standard definitions of many types of development and laying down formulae for specific zones, the template demands a high level of conformity. In some circumstances, it might be argued as a means to erode creativity in local plan making (Kelly and Smith, 2008). This phenomenon of sameness is reflected in suburban sprawl. While building fashion might change, uniformity tends to reign across new housing landscapes.

SEPPs and LEPs tend to provide regulatory rather than incentive clauses. Nevertheless, they can reach far beyond regulating uses such as buildings, subdivision, industry and mines. Under the EPAA, an EPI makes provisions for, *inter alia*:

(a) protecting, improving or utilising, to the best advantage, the environment,
(b) controlling (whether by the imposing of development standards or otherwise) development . . .
(e) protecting or preserving trees or vegetation,
(e1) protecting and conserving native animals and plants, including threatened species, populations and ecological communities, and their habitats,
(f) controlling any act, matter or thing for or with respect to which provision may be made under paragraph (a) or (e) . . .[3]

It is clear that an LEP may regulate front and back yards. For instance, it may require consent for the removal of one or more specified trees. Alternatively, the decision-maker may approve residential development subject to certain trees being retained. A council might even demand that a proposal be redesigned in order to retain identified vegetation, or replace trees to be removed with more suitable native species.

There is concern, however, as mentioned before, that contemporary detached dwellings contain smaller yards. The modern home is often enormous. Plot coverage of dwellings across urban Australia has expanded substantially with garden space 'almost completely covered by larger dwellings' (Hall, 2007, 26). This has captured recent media interest, with a *Sydney Morning Herald* front page article reporting that the size of Australian homes is 'overtaking those in the US as the world's biggest' with Sydney's 'new free-standing houses typically spanning 263 square metres' (Martin, 2009, 1; see also Curtin, 2009; Frew, 2009). This is backed by legislative change. Under the 2008 NSW 'Housing Code',[4] allotments of between 450 and 600 square metres can accommodate up to 50 per cent building coverage (NSW Department of Planning, 2008, 8–12). But this excludes driveways, verandas, terraces, cabanas and even swimming pools and spas, leaving little room for backyard cricket or planting native trees. Furthermore, these types of developments need not undergo environmental assessment; instead, there is a straightforward 'tick the box' approach to obtain permission. They fall into the category of 'complying development' which may be handled by private certifiers.[5] The state government's focus is on efficiency, quick approvals, rapid development and minimum environmental intervention by local government. The resultant urban sprawl is therefore advancing across far-flung suburbia. It is not only encouraged but expected.

9.3 Amenity

9.3.1 The meaning of amenity

Protection or enhancement of amenity is, even in small yards, immediately relevant. Over four decades ago, Wilcox (1967, 361) described amenity as the 'hardest worked word in planning language'. Yet amenity is an entrenched concept in town planning practice (Cullingworth, 1967; McAuslan, 1980). This is illustrated by the landmark Housing, Town Planning etc Act 1909 (UK) which empowered local authorities to formulate schemes for areas in the course of development under the 'general object' to 'secur[e] proper sanitary conditions, amenity and convenience in connection with the laying out and use of the land'.[6] In presenting the bill, the president of the local government board triumphed that Britain's first planning statute would secure:

> the home healthy, the house beautiful, the town pleasant, the city dignified and the suburb salubrious. It seeks and hopes to secure more houses, better houses, prettier streets, so that the character of a great people in towns and cities and villages can be still further improved and strengthened . . ..[7]

NSW followed suit. The CCPS ingrained amenity protection into planning regulation by requiring councils when determining development applications to 'take into consideration', *inter alia*:

> the existing and likely future amenity of the neighbourhood including the question whether the proposal is likely to cause injury to such amenity including injury due to the emission of noise, vibration, smell,

fumes, smoke, vapour, steam, soot, ash, dust, grit, oil, waste water, waste products or otherwise.[8]

This factor was reproduced in subsequent Planning Scheme Ordinances (PSOs) across NSW. When the EPAA commenced operation in 1980 with matters for consideration set into the legislation rather than in individual plans, the then provision included 'the existing and likely future amenity of the neighbourhood'.[9] In contrast, the original paragraph in the CCPS was far more specific regarding the types of potential 'injury' to neighbourhood amenity. Its focus was on neighbourhood pollution, particularly from industrial uses, harking back to the post-industrial origins of planning law. But in 1955, the judiciary made it clear that 'injury' to amenity extended beyond those causes expressly listed to include, *inter alia*, impact on the *visual* environment.[10] This notion became ensconced in development control law. For instance, in a frequently cited judgement, Sugerman J regarded amenity as crucial in terms of the visual effect of a service station on a residential neighbourhood, stating that:

> to break up a line of residences with their lawns and gardens by the interposition of a service station building with its paved yard, equipment of petrol pumps and other accessories, and daily congestion of parked vehicles is to detract from the pleasurable appearance of a neighbour- hood in the eyes both of residents and passers-by.[11]

As the planning system moved onwards, councils directed their energy to 'the protection of local amenity, usually residential amenity' (Harrison, 1988, 27; see also Stein, 2008). In the late 1990s, amendments to EPAA scaled back the original 27 matters for consideration to five.[12] Whilst this deleted any specific reference to amenity, the judiciary has made it clear that the short- ened list was not exhaustive. Amenity is now a frequently raised issue before the specialist Land and Environment Court (LEC) of NSW, especially in merits appeal cases (Kelly, 2006). In a recent case,[13] for example, which involved construction of telecommunications equipment on the roof of a club, the LEC described amenity as 'wide and flexible'.

Smith (1974, 260) describes amenity as encompassing 'environmental health, pleasantness and civic beauty'. The visual context reflects the facts that most people rely on sight more than any other sense (Tuan, 1974). Amenity is, therefore, of major significance for suburban residents and passers-by who enjoy green front yards. Taylor (1999, 59) refers to 'picturesque suburbia as a national icon' while Herzog (1995) highlights 'tended nature' as especially popular. Such commentary underlines the sheer attractiveness of leafy gar- dens as opposed to unrelenting concrete. But it does not demand undisturbed ecological systems. Dawson (1990, 138) astutely observes 'nature' as:

> highly desired in the urban garden . . . [yet] has its own architecture, one far more complicated and diverse than human architecture. The

architecture of nature is ecology. Garden ecology is the application of this to gardens.

Since smaller gardens are becoming the norm in new residential estates and suburban infill, their modification to 'nature' is turning out to be more extreme.

9.3.2 Amenity, gardens and the law

When originally enacted, the EPAA included, *inter alia*, objects that encouraged the 'proper management, development and conservation of natural and artificial resources . . . for the purposes of promoting the social and economic welfare of the community and a better environment' and the 'protection of the environment'.[14] The second phrase was later expanded to include 'the protection and conservation of native animals and plants including threatened species, populations, ecological communities and their habitats'.[15] While not explicitly addressing amenity, the concept was nonetheless embedded within these provisions. The EPAA also included the object of encouraging the 'promotion and co-ordination of the orderly and economic use and development of land'.[16] A provision recognizing a need for affordable housing was also later added.[17] As land is often cheaper at the outskirts of major cities, these competing objectives become especially relevant at the urban periphery particularly where remnant native vegetation is threatened.

Impact on amenity is often crucial in determining applications for development (Stein, 2008). For instance, disputes over proposals such as small acreage subdivision on vegetated land or erection of dwelling houses on bushy sideslopes may arise. In the suburban context, in addition to public parklands and street verges, residential gardens play a key role especially via front yards which are more visible. Because amenity is a subjective concept, it reflects personal preferences and community culture. Landholders may prefer to change their neighbourhood landscapes with exotic trees rather than maintain what might appear as tedious scrub (Kelly, 2006). For instance, assemblages of the remnant Cumberland Plain Woodland in western Sydney, a critically endangered ecological community listed under the *Threatened Species Conservation Act 1995* (NSW) (TSCA), might be regarded as drab with landholders preferring colourful exotic species such as the South American jacaranda or a variety of tropical palms. Of course, such temptations are visible across all suburbs. Attention is now being paid to integrating private residential land with local and 'safe' bushland – i.e. indigenous vegetation recognized as not only ecologically appropriate to the neighbourhood but also meeting security concerns. Most residents will want to avoid prickly plants and those perceived to attract venomous spiders. In expanding Sydney, this leads to separate shrubs rather than towering trees and replication of native mosaics. This fits in with the onslaught of immense houses and smaller yards. Residents are now more likely to seek recreational benefits in their local parks (Halkett, 1976; Syme *et al.*, 2001; Head and Muir, 2007). A bigger issue, however, as will be seen later, is bushfire reaching suburbia.

Amenity protection can be expressed in individual LEPs. A ready example is Liverpool LEP 2008 at Sydney's edge, partially in the South-West Growth Centre. The instrument was a pioneer that follows the LEP template. Unsurprisingly, amenity is undefined. Yet the notion is addressed in a main objective to 'maintain suitable amenity and offer a variety of quality lifestyle opportunities to a diverse population'.[18] Further examples relate to temporary use of land, various zonal objectives especially in relation to residential zones, minimum subdivision size and foreshore building lines. Perhaps the most interesting is the template clause relating to '[p]reservation of trees or vegetation' that requires permission for tree removal or damage. This item derives from the 'tree preservation order' (TPO) item in the CCPS and many subsequent local instruments. Indeed, such clauses originate from British planning ordinances (Cullingworth, 1967). As Liverpool LEP 2008 demonstrates, amenity remains a vital component of the statutory planning jigsaw. But while the TPO template clause provides the latest regulatory tool for protecting amenity, its adoption by councils is totally optional.

Interference with vegetation that is contrary to a TPO may attract criminal action. For instance, the LEC has emphasized that 'breach of a tree preservation order is a serious offence'.[19] In this case, the defendant was found guilty of removing two trees in the western suburb of Westmead, including a Queensland Fire Wheel twelve metres in height which was reported as providing 'existing amenity' and 'colour, shade and screening'.[20] The defendant admitted guilt and was fined $A15,000. This and many other judgements illustrate how amenity is cemented in planning law. It relates directly to the appearance of suburban gardens. In contrast, biodiversity conservation and protection from bushfire represent far more recent concepts.

9.4 Biodiversity conservation

9.4.1 The notion of biodiversity

Biodiversity is a different model altogether. It is based on science, representing a far more modern phenomenon within planning law and municipal policy. It is also extremely complex. The National Strategy for the Conservation of Australia's Biological Diversity defines biodiversity as 'the variety of all life forms – the different plants, animals and micro-organisms, the genes they contain and the ecosystems of which they form part' (Commonwealth of Australia, 1996, 1). Accordingly, it embraces the inconspicuous, the bleak and the malodorous: matters that exist well beyond the amenity spectrum.

Whilst the precise origins of the term are arguable (Adam, 2009), Jeffery (1997, 4–5) refers to a 'snappy abbreviation' composed by the co-director of the 1986 American 'National Forum for BioDiversity' who recognized references to 'biological diversity' in earlier scientific papers. The term has since become far more fashionable, often found in tourist brochures and newspaper articles. In a recent weekly gardening column from the *Sydney Morning Herald*, the author warns readers that because Australia has 'one of the worst records for loss of biodiversity', '[g]ardeners can be of great help to native birds and animals by cultivating indigenous plants to provide green corridors' (Maddocks, 2009, 25). The essence is no different to Beatley's (2000) paper on retaining biodiversity in American back yards, even in small gardens. As noted by Hall (2007, 27), '[p]rivate gardens exhibit a high degree of biodiversity'. All this reflects the fact that biodiversity conservation need not be restricted to the pristine. On the other hand, minimal suburban yards might be more of a museum than a working green environment. The ambush of non-indigenous plants from gardens into nearby bush raises problems (Zagorski *et al.*, 2004), in addition to severe habitat modification. McKinney (2006, 248) refers to urbanization as 'one of the most homo-genizing activities of all' due to its 'exceptionally uniform nature'. These factors suggest that amenity can be an anathema to biodiversity conservation and that the notion of 'garden ecology' raised earlier must not be disguised as a solely scientific-based approach to conservation.

Biodiversity protection in Australia is crucial at the regional, national and global levels. Its international dimension is incorporated in the Convention on Biological Diversity, signed by many countries, including Australia, at the 1992 Rio Earth Summit. Australia's ratification was vital given its 'mega-diverse' nature. It is 'geographically more isolated' than other countries with rich biodiversity and the 'only one . . . predominantly in the temperate region' (New, 2000, 23). Possingham (2008) adds that a huge number of Australian species are endemic. Yet conservation biologists, policy-makers and environ-mental lawyers must consider well beyond listed species to other aspects, such as ecological assemblages. As Adam warns (2009, 19), 'the 'big picture' approach is not being adopted and attention and resources are still on listed species'. It is here where the complexity intensifies. Specifying the border lines of ecological communities is scarcely easy. Boundaries are far from static.

9.4.2 Biodiversity conservation law

All spheres of government in Australia are involved in biodiversity law and policy. Strategic documents have been designed at each level, including local government (Australian Local Government Association and Biological Diversity Advisory Council, 1999). At the top of the list of conservation objectives in this document is 'high economic returns through tourism and increased land values, due to scenic and amenity values' (at 35). This seems inconsistent with the scientific demands of biodiversity conservation.

In NSW, much biodiversity conservation is primarily delivered through the TSCA. Supporting this, many councils have prepared their own voluntary biodiversity policies which rely on sufficient monetary resources, political backing and staff expertise. They do not carry statutory force but instead may offer specialist information or provide incentives such as free or subsidized seedlings. Such programs are very different from regulatory control, with documents providing useful educational material for local citizens. In its own 'biodiversity strategy', Liverpool City offers an apparent scientific approach with tables, technical information and a series of sub-issues such as suggested strategies, proposed actions, recommended policies (e.g. conservation targets, corridors and connectivity), tools and resources plus detailed maps (Liverpool City Council and Ecological Australia, 2003). The council relied heavily on consultant expertise here. The document reflects a non-statutory commitment to biodiversity conservation with implementation dependent on political and financial support. Nevertheless, it is far superior to those situations where a council has no such policy at all. Notably, Liverpool City Council draws attention to the Cumberland Plain Woodland which, as noted earlier, is a listed threatened ecological community under the TSCA. Upon European settlement, this ecological community covered over 120,000 hectares (NSW National Parks and Wildlife Service, 2004). It has since been decimated to 9 per cent of its original extent, with a further 14 per cent remaining as scattered trees. These include small but important patches in residential yards. What advantages might the planning system present here?

The TSCA piggybacks on the EPAA by demanding special considerations and requirements for listed threatened species, populations, and ecological communities. If any of these items or their habitats will or may be adversely affected by a proposed EPI, including an LEP, then consultation with the responsible state agency for threatened species is required (currently, the Department of Environment and Climate Change and Water (DECCW)).[21] While such consultation potentially enables components of biodiversity to be included in strategic land-use decisions, there is no obligation on the operator of the EPI to apply the advice received. Additionally, there are no prescribed guidelines for how threatened species are to be conserved in urban planning decisions.

Threatened species considerations for development proposals are, however, more specific. All development applications must undergo a 'seven point test' to determine if a proposal is likely to have significant impact on a listed threatened species (e.g. Koala), population (e.g. Little Penguin in the Manly Point Area) or ecological community (e.g. Cumberland Plain Woodland), or their habitats.[22] If it is decided that the impact is significant, a 'species impact statement' must accompany the application and DECCW has a 'right of veto' over the development.[23] This might relate to, for example, a residential subdivision in a bushy acreage in Sydney's West or erection of a building where a listed ecological community exists. Assessment is mostly carried out by councils or consultants on their behalf. These requirements may deter needless ecological damage. Detailed 'seven point test' assessments can encourage designs that minimize adverse ecological effects. However, even if the decision-maker concludes that the environmental impact will be ecologically devastating, approval is still possible (Kelly and Farrier, 1996). Local and state politics may demand more suburban development leading to informed habitat destruction. Indeed, there has been increasing concern that the narrow focus on threatened species has taken the spotlight away from wider biodiversity and landscape-scale conservation issues, giving rise to *ad hoc* decision making and evaluation at the very end of the planning process (Bubna-Litic, 2008). As noted by Riddell (2005, 446), 'the balance is skewed strongly in favour of development and economic growth'.

The TSCA has undertaken further amendments, including potential for 'biocertification' of certain areas to avoid the 'seven point test'. Detailed consideration of these mechanisms is beyond the limits of this chapter. It is worthy of note, however, that the Growth Centres SEPP for Western Sydney received biocertification by ministerial order in late 2007 and was supported through direct amendments to the TSCA in 2008. Accordingly, the state government has endorsed a high-level approach to resolve urban development and biodiversity conflicts by using a process of certification, and incorporating the use of conservation offsets, to bypass the need for individual threatened species assessments at the property scale. This illustrates how biodiversity law is volatile and contemporary. It is now a strong element in the planning system but relates to process and procedure rather than final outcomes.

While councils have opportunity to deal with biodiversity conservation, it appears that this is largely restricted to policy and strategic instruments rather than local statutory instruments, namely LEPs. Liverpool LEP 2008 includes only one reference to biodiversity ensuring that a proposal on land recognized as 'high biodiversity significance' cannot fall within the list of 'exempt developments' which otherwise enables proposals to escape the development control process. Further, clauses for nature conservation relate to certain mechanisms such as the 'large lot residential' zone and the modern TPO-derived provision on 'preservation of trees and vegetation'. As noted earlier, this LEP adheres to the LEP template. TPO provisions within any LEP, as argued by Kelly (2006), usually confuse both amenity enhancement and biodiversity protection. More clarity is needed.

Conservation of biodiversity is a difficult objective for ill-equipped and conservative councils. While amenity is of neighbourhood concern, biodiversity conservation is an international concept. Even with more funding and direction, some councillors and officers may restrict their attention to rate-payer issues (Kelly and Stoianoff, 2006). Even though they must confront biodiversity conservation under the law, they may prefer to address less controversial tactics such as convening helpful workshops on monitoring native vegetation in private gardens. Bushfire, in contrast, attracts immediate attention from all directions.

9.5 Bushfires

9.5.1 *The impact of bushfire in suburban Australia*

Unlike the other two topics, wildfire, or bushfire as it is known in Australia, can immediately involve loss of human life and property. It therefore has an imperative role to play in development at the urban edge. It also attracts more media attention than restoring fragments of Cumberland Plain Woodland. The social and economic implications are greatest in the densely populated areas of southern Australia (Russell-Smith *et al.*, 2007). This is reflected in major events during the past fifty years, the most recent being the Victorian bushfires of 7 February 2009 which resulted in the tragic death of 173 people with around 2,000 houses destroyed (Victoria Bushfire Reconstruction and Recovery Authority, 2009).

While the Victorian bushfires affected rural townships, in other places bushfires have reached the suburbs. In 2003, a firestorm penetrated several suburbs of Canberra, Australia's capital city, resulting in four deaths and the loss of 530 houses (Odger *et al.*, 2003). Other bushfires in NSW during December 2001 and January 2002 destroyed 109 homes including peri-urban areas around Sydney (Little, 2002). In 1994, bushfires unexpectedly leapt across a valley in southern Sydney to destroy 101 homes with one person dying from heat and smoke as she attempted to reach her swimming pool (Cockerill, 1994; Cheney 1995). The area burnt was only 476 ha, indicating that even 'small' fires can have major social and economic consequence when located in close proximity to urban areas (Cheney 1995; Gill and Moore, 1998). This short list reflects the particularly high risk occurring at the edges of Australia's towns or cities where residences adjoin bushland (McAneney *et al.*, 2009).

Bushfires occur when there is a favourable combination of weather, fuel and ignition source (Cheney 1979). As stressed by Gillen (2005), Sydney is especially prone to bushfire; it enjoys a subtropical climate with 'summer temperatures frequently reaching the high 30s Centigrade and bringing low humidity and warm conditions' (Gillen, 2005, 466). Sydney is also characterized by an urban landscape where extensive areas of fire-prone native vegetation occurs adjacent to and within urban areas. This is a legacy influenced by a sandstone geology which restricted the direction of early settlement due to the presence of steep sideslopes where nutrient-deficient skeletal soils offered scant agricultural promise (Haworth, 2003). Many of these forested areas now exist as national parks or council-managed reserves, creating the situation where the bushfire hazard often resides in nearby public ownership. Remote sensing analysis has identified that about 189,000 (6.6 per cent) of residential addresses within Sydney and its hinterlands fall within 80 m of extensive bushland and are therefore at greatest risk from bushfire (Chen, 2005). In fact, due to its many fingers of bushland, Sydney and its hinterland account for more than half (56 per cent) of the national residential addresses falling within 80m or 130m of bushland (Chen, 2005).[24] Gillen (2005, 466) also observes that Sydney 'is located in the zone of highest bushfire frequency in Australia'. This dovetails with recent research which has found that 33–50 per cent of all vegetation fires occur in or around a capital city, with the greatest concentrations being at the bushland–urban interface (Bryant, 2008).

9.5.2 Dealing with bushfire in the urban periphery

Historically, there have been two approaches to mitigating building and property destruction from bushfire. One has been to manage vegetation across the landscape whereas the other has focused on increasing building resistance to bushfire attack (Ramsay *et al.*, 1995). Recent attention has centred on means to integrate both approaches to both maximize property protection and minimize environmental impacts such as on biodiversity (Ramsay *et al.*, 1995; Bradstock and Gill, 2001; Bradstock, 2003). Strategies that manipulate vegetation at the bushland edge have been advocated as potentially providing the optimal solution to the 'bushfire problem' (Bradstock and Gill, 2001; Bradstock, 2003).

One of the primary means of achieving this is through the creation and management of a permanent fuel-reduced area between urban dwellings and the bush (Jasper, 1999; Bradstock and Gill, 2001; Leonard, 2003). This responsibility has traditionally fallen on land managers rather than being accommodated in the footprint of proposed development. It has forced land managers and owners of bushland to reduce hazards on their own land for the benefit of neighbours. Unfortunately, when coupled with demands and expectation from developers to maximize lot yields, especially with larger dwellings, this unwritten expectation has allowed urban development to sprawl along the bushland edge with limited bushfire protection measures. While planning policy and legislative change during the past decade have made new development more accountable, opportunity to address *previous* poor design is limited. Protection of existing development remains largely reliant on modifying adjoining bushland through bushfire hazard reduction activities rather than controls that encourage or require residents to upgrade buildings and/or adjacent yards.

Implementation of bushfire hazard reduction activities can encounter substantial difficulty. It is often resource intensive due to local terrain and vegetation constraints. Action is reliant on 'windows of opportunity' of favourable weather conditions (Conroy, 1996; Gill and Moore, 1998; Little, 2003). Put simply, bushfire hazard reduction measures cannot replace effective urban planning designs such as 'emergency access routes, water supply requirements, lot depth and configuration, setbacks and building construction materials' delivered through effective urban design (Little, 2003, 32). This leads directly to local government's planning function in both plan making and implementation.

Past approaches to urban design give some insight to the problem at the urban–bushland edge. Prior to 2002, bushfire protection provisions were mostly advisory rather than mandated into the NSW planning system (Conway and Lim, 2002; Little, 2002). This inadvertently gave rise to development designs which maximized lot yield and economic returns to developers who positioned housing lots on either side of central access roads. In bushland settings, this enabled front yards to face the access road while the back yards adjoined the bush. Such properties, which are still common, are often characterized by limited setback distances from flammable bush-land due to lot depth and poor integration with key infrastructure, such as water supply, access roads and fire trails. Back yards are also less conspic-uous, enabling dumping of rubbish into the bush and disturbing biodiversity. Such actions can directly increase available fuel load during a bushfire event as well as increasing the propensity for weed invasion. Council backing and/or landholder preferences for amenity or biodiversity objectives can also result in increased risk to dwellings if bushfire protection considerations are overlooked.

9.5.3 Changes to bushfire law and effects on private open space

Following large bushfires in NSW during 2001–2002, major statutory change was introduced to improve the role of development planning and assessment procedures in reducing urban vulnerability to bushfire. It expanded the role of the NSW Rural Fire Service (RFS) in assessing critical development decisions throughout bushfire prone areas. These reforms amended the *Rural Fires Act 1997* (NSW) (RFA) and the EPAA (Little, 2002). A new guideline, 'Planning for Bushfire Protection', was prepared jointly by the RFS and the state planning agency in 2001 to deal with land use planning, subdivi-sion proposals and applications for building in designated bushfire-prone areas (NSW Rural Fire Service and PlanningNSW, 2001). The reforms explicitly required councils to map bushfire-prone areas on a bushfire-prone land map,[25] thereby extending the scope of planning further. The revisions increased obligations on councils to comply with certain provisions (e.g. access arrangements, asset protection zones (APZs), water supply require-ments) and consult with the RFS in drafting their LEPs.[26] Councils were also required to refer subdivision applications to the RFS for approval,[27] and ensure that new building proposals in bushfire-prone areas complied with the said guideline or consulted with the RFS regarding alternative measures.[28]

The urban designs advocated through 'Planning for Bushfire Protection' (NSW Rural Fire Service and PlanningNSW, 2001) and its more recent replacement, 'Planning for Bush Fire Protection' (NSW Rural Fire Service, 2006) have strongly influenced urban design at the bushland edge. Sub-division developments are now required to provide a two-way perimeter road between residential lots and the bush. In certain instances, perimeter fire trails are allowed. These form part of the fuel-managed area known as the APZ. The other part of the APZ is made up of the future yards of subdivided lots.

Perimeter roads not only assist in separating residential dwellings from the bush but provide firefighters with easier access to protect buildings and a clear control line for prescribed burning or emergency backburning operations. This ultimately facilitates a more effective and efficient use of resources by enabling buildings to be protected from one access point via the road. Such designs give rise to front gardens that face the perimeter road but are separated from the bush by the road. The back yards face other houses and yards in the block but not the bush. As front yards are more visible, residents are less likely to keep such areas unkempt. Landholders may maintain their gardens with lawn, rockeries and the like, enhancing amenity and minimizing bushfire risk.

Some aspects of design in the built environment, such as management of gardens that influence the susceptibility or resilience of a dwelling to bushfire attack, fall under the responsibility of private individuals (Bradstock, 2003). The presence and density of vegetation is known to affect building survival with Cyprus pines, pencil pines and golden cypress being particularly hazardous (Blanchi *et al.*, 2006). Landscaping priorities such as selection and location of plants, manipulation of landform, siting and design of walls, fences, driveways, and paths, and even the types of mulches, can influence susceptibility to bushfire attack (Ramsay *et al.*, 1995; Ramsay and Rudolph, 2003). While such strategies are generally equated with decreasing herbage biomass, appropriately sited low-flammable vegetation can assist in creating barrier effects that trap embers, and decrease radiant heat and wind velocity (Leonard, 2003; Ramsay and Rudolph, 2003, Blanchi *et al.*, 2006). However, the use of vegetation in this way often requires large lots. It is generally more available in rural residential settings rather than suburban lots. Traditionally, landscaping matters have been largely unregulated and left to the landholder's choice. Accordingly, while urban planning may lead the way in a strategic sense to reduce risks to property in bushfire prone areas, owners of land must take care in choosing plants for their gardens and maintaining their allotments. Reliance on governmental regulatory control alone is insufficient to effectively safeguard property from bushfire. Overregulation is usually politically unpopular. This leaves landscaping largely reliant on visionary consent conditions and the knowledge, skills and willingness of landholders to manage their gardens to minimize bushfire risk.

In terms of LEPs, provisions relating to bushfire are generally common at the urban periphery. But this may be waning. The LEP template is sparse regarding references to bushfire. The only clause that stands out reads that 'bush fire hazard reduction may be carried out on any land without consent'.[29] This provision is mandatory, as is reflected by Liverpool LEP 2008. Additional clauses in the LEP relate to its overall aims, complying development and the need for a supporting 'development control plan' (DCP)[30] for urban release lands to address bushfire. This involves wider consideration of bushfire issues than afforded by the template. In terms of LEPs, provisions relating to bushfire under the template sit in stark contrast to those under the relevant ministerial direction.[31] The relationship between this direction, the template and DCPs with regard to bushfire protection warrants further exploration given the changing legislative terrain of land use planning and evolving garden designs.

9.6 Conclusions

This chapter presents a conundrum. It approaches three very different issues relating to front and back yards, needing to be integrated in a manner that serves both neighbourhood and global interests. Protection of amenity deals essentially with local visual attractiveness. It is largely subjective, involving establishment and maintenance of green charm. In contrast, biodiversity conservation engages international and national influences translated to regional and local levels. Biodiversity also relates to environmental impact assessment and threatened species law, embracing protection of native vegetation and habitat. Based on science rather than local concerns, it may compete directly against amenity. However, there are opportunities for urban amenity to work hand-in-hand with biodiversity such as when landscape design can incorporate strategies that engage appropriate locally endemic species. But at the bushland–urban interface, protection from bushfire cannot be underplayed. It attracts strong media interest and imme-diate government responses, especially when there is loss of human life. Effective protection from bushfire involves vegetation modification, land-scaping and setback distances between houses and the bush. It therefore has a significant role in influencing the design and management of front and back yards at the urban edge.

Several key factors must be recognized here. First, the three aspects illustrate an evolving planning framework based on societal expectations. Arguably, amenity protection is losing out. Larger houses and smaller gardens leave less opportunity for enhancing amenity with expanded concrete driveways and triple garages. On the other hand, perhaps the expec-tations of suburban amenity are changing. The traditional role of amenity enhancement contrasts with biodiversity conservation and bushfire pro-tection, which are more recent and essential ingredients in modern local planning.

Second, many planners must broaden their professional understanding of bushfire issues. They must embrace the need to reform streetscapes and influence private garden design at the urban periphery where new develop-ment abuts bushland. Councils should confront the need for perimeter roads to prevent back yards from directly fronting bushland. While this may diminish direct bushland experience, it sends a clear message that native vegetation represents other values. It requires distinct management tech-niques from maintaining manicured gardens.

Third, many councils are already involved in assisting local residents on appropriate plants for their gardens, including free or subsidized appropriate saplings. Councils may, for instance, hold public fora and courses on local vegetation and landscape design. Teaching at local schools and establishing special plots of land to introduce and maintain preferred native species is possible. Matters such as local floral symbols and bushfire history are worthwhile. The advantage here is that it moves away from the notion of an over-regulatory 'nanny state'. On occasions, hard decisions must still be made such as refusing consent to remove safe trees in the back yard. But a well-informed community will be in a better position to tolerate this.

A further issue is reinforcement of *regional* planning. This is a central plank of the NSW Department of Planning, with its regional strategies and growth centres policy. Perhaps a softer approach to reasonable changes to the LEP template would assist further. The test is to transcribe government mandates into practical and functional outcomes that conserve high value natural environments and provide safe, scenic and affordable suburbs. Delivery of this balance at the urban periphery remains a major challenge for local government at the outskirts of Sydney. More attention to the design of sustainable front and back yards is vital.

Notes

1 See State Environmental Planning Policy (Sydney Regional Growth Centres) 2006.
2 *Environmental Planning and Assessment Act 1979* (NSW), s 33A.
3 Section 26.
4 See NSW State Environmental Planning Policy (Exempt and Complying Development Codes) 2008.
5 *Environmental Planning and Assessment Act 1979* (NSW), ss 84–87.
6 *Housing, Town Planning etc Act 1909* (UK), s 54(1).
7 Hansard, 4th series, Vol clxxxviii, 960–61, 12 May 1908 (UK).
8 Cumberland County Planning Scheme, cl 27(e).
9 *Environmental Planning and Assessment Act 1979* (NSW), former s 90(1)(o).
10 *Tooth & Co v Parramatta City Council* (1955) 20 LGR (NSW) 60 at 75–76.
11 See *Vacuum Oil Co Pty Ltd v Ashfield Municipal Council* (1956) 2 LGRA 8 at 12.
12 *Environmental Planning and Assessment Act 1979* (NSW), s 79C(1).
13 *Telstra Corp Ltd v Hornsby Shire Council* (2006) 146 LGERA 10 at 51.
14 *Environmental Planning and Assessment Act 1979* (NSW), ss 5(a)(i), (vi).
15 Section 5(a)(vi).
16 Section 5(a)(ii).
17 Section 5(a)(viii).
18 Liverpool Local Environmental Plan 2008, cl 1.2(2)(c).
19 *Holroyd CC v Skyton Developments Pty Ltd* (2002) 119 LGERA 225 at 229, per Cowdroy J.
20 At 227.
21 *Environmental Planning and Assessment Act 1979* (NSW), s 34A.
22 Section 5A.
23 Section 79B.

24 This is based on an analysis of Australia's major capital cities.
25 *Environmental Planning and Assessment Act 1979* (NSW), s 146.
26 Ministerial Direction G20 – Planning for Bushfire Protection, issued under the *Environmental Planning and Assessment Act 1979* (NSW), s 117(2). This has since been replaced by Ministerial Direction 4.4, also known as Planning for Bushfire Protection.
27 See *Environmental Planning and Assessment Act 1979* (NSW), s 91(1); Rural Fires Act 1997 (NSW), s 100B(1).
28 *Environmental Planning and Assessment Act 1979* (NSW), Section 79BA.
29 Liverpool Local Environmental Plan 2008, cl 55.
30 See *Environmental Planning and Assessment Act 1979* (NSW), ss 74B–74F. Unlike an environmental planning instrument, a development control plan (DCP) holds persuasive rather than statutory force. Relevant provisions in a DCP that apply to a parcel of land subject to a development application, however, must be taken into account by the decision maker.
31 Ministerial Direction 4.4 – Planning for Bushfire Protection, made under *Environmental Planning and Assessment Act 1979* (NSW), s 117(2).

References

Adam, P. (2009), 'Ecological communities – the context for biodiversity conservation or a source of confusion?', *The Australasian Journal of Natural Resources Law and Policy*, Vol. 13 No. 21, pp. 7–59.
Alexander, I. (2000), 'The post-war city', in Hamnett, S. and Freestone R. (Eds.), *The Australian Metropolis: A Planning History*, E & FN Spon, London, pp. 98–112.
Australian Local Government Association and Biological Diversity Advisory Council (1999), *National Local Government Biodiversity Strategy*, Australian Local Government Association, Canberra, ACT.
Beatley, T. (2000), 'Preserving biodiversity: challenges for planners', *Journal of the American Planning Association*, Vol. 66 No. 1, pp. 5–20.
Blanchi, R., Leonard, J. E. and Leicester R. H. (2006), 'Lessons learnt from post bushfire surveys at the urban interface in Australia', in *Proceedings of the V International Conference on Forest Fire Research*, 27–30 November, Coimbra, Portugal, available at http://www.bushfirecrc.com/research/downloads/Blanchi-Coimbra-final-paper.pdf (accessed 15 November 2009).
Bradstock, R. A. and Gill, A. M. (2001), 'Living with fire and biodiversity at the urban edge: in search of a sustainable solution to the human protection problem in southern Australia', *Journal of Mediterranean Ecology*, Vol. 2, pp. 179–195.
Bradstock, R. A. (2003), 'Protection of people and property: towards an integrated risk management model', in Carey, G., Lindenmayer, D. and Dovers, S. (Eds.), *Australia Burning: Fire Ecology, Policy and Management Issues*, CSIRO Publishing, Collingwood, Victoria, pp. 119–123.
Bryant, C. (2008), 'Deliberately lit vegetation fires in Australia', *Trends and Issues in Crime and Criminal Justice,* No. 350, available at: http://www.aic.gov.au/documents/5/6/4/%7B564041DC-F1B7-4BA3-930F-ED7868B2A37C%7Dtandi 350.pdf (accessed 15 November 2009).
Bubna-Litic, K. (2008), 'Ten years of Threatened Species legislation in NSW – what are the lessons?', in Jeffery, M. I., Firestone, J. and Bubna-Litic, K. (Eds.), *Biodiversity Conservation, Law and Livelihoods: Bridging the North-South Divide*, Cambridge University Press, Cambridge, pp. 265–279.

Chen, K. (2005), 'Counting bushfire-prone addresses in the Greater Sydney Region', in *Planning for Natural Hazards – How Can we Mitigate the Impacts? Symposium*, 2–5 February, 2005, GeoQuest Research Centre, University of Wollongong, NSW, Australia, pp. 45–56.

Cheney, N. P. (1979), 'Bushfire disasters in Australia 1945–1975' in Heathcote, R. L. and Thom, B. G. (Eds.), *Natural hazards in Australia: proceedings of a symposium sponsored by Australian Academy of Science Institute of Australian Geographers Academy of the Social Sciences in Australia*, Australian Academy of Science, Canberra, ACT, pp. 72–93.

Cheney, N. P. (1995), 'Bushfires – an integral part of the Australian environment', in *Year Book Australia, 1995*, Australian Bureau of Statistics, Canberra (ABS Catalogue No. 1301.0), available at: http://www.abs.gov.au/Ausstats/abs@.nsf/Lookup/6C98BB75496A5AD1CA2569DE00267E48 (accessed 18 November 2009).

Cockerill, I. (1994), *The Wildfires of 1994*, Weldon Kids, Willoughby, NSW.

Commonwealth of Australia (1996), *National Strategy for the Conservation of Australia's Biological Diversity*, Australian Government Publishing Service, Canberra, ACT.

Conroy, R. J. (1996), 'To burn or not to burn? A description of the history, nature and management of bushfires within Ku-ring-gai Chase National Park', *Proceedings of the Linnean Society of New South Wales*, Vol. 116, pp.79–95.

Conway, A. and Lim, L. (2002), 'The recent bushfire crisis in NSW. Where to from here?', *Local Government Law Journal*, Vol. 7 No. 3, pp. 169–176.

Cullingworth, J. (1967), *Town and Country Planning*, George Allen and Unwin, London.

Curtin, J. (2009), 'How the mcmansion super-sized the suburbs', *Sydney Morning Herald*, 5 December, p. 7.

Davison, G. (1994), 'The past and future of the Australian suburb', in Johnson, L. (Ed.), *Suburban Dreaming: An Interdisciplinary Approach to Australian Cities*, Deakin University Press, Geelong, Victoria, pp. 99–113.

Dawson, K. J. (1990), 'Nature in the urban garden', in Francis, M. and Hester, R. T. (Eds.), *The Meaning of Gardens: Idea, Place and Action*, MIT Press, Cambridge, Massachusetts, pp. 138–143.

Frew, W. (2009), 'Mansions welcome under new code', *Sydney Morning Herald*, 1 December, p. 5.

Gill, A. M. and Moore, P. H. R. (1998), 'Big versus small fires: the bushfires of greater Sydney', January 1994', in Moreno, J. M. (Ed.), *Large Forest Fires*, Backhuys, Leiden, The Netherlands, pp. 49–68.

Gillen, M. (2005), 'Urban vulnerability in Sydney: policy and institutional ambiguities in bushfire protection', *Urban Policy and Research*, Vol. 23 No.4, pp. 465–476.

Halkett, I. P. B. (1976), *The Quarter Acre Block: The Use of Suburban Gardens*, Australian Institute of Urban Studies, Canberra, ACT.

Hall, T. (2007), *Where Have All the Gardens Gone? An Investigation into the Disappearance of Back Yards in the Newer Australian Suburb*, Urban Research Program, Griffith University, Brisbane, Qld.

Harrison, P. (1988), 'Urban planning and urban issues: 1951–72', *Australian Planner*, Vol. 26 No. 3, pp. 26–27.

Haworth, R. J. (2003), 'The shaping of Sydney by its urban geology', *Quaternary International*, Vol. 103, pp. 41–55.

Head, L. and Muir, P. (2007), *Backyard: Nature and Culture in Suburban Australia*, University of Wollongong Press, Wollongong, NSW.

Herzog, T. (1995), 'A cognitive analysis of preference for urban nature', in Sanha, A. (Ed.), *Readings in Environmental Psychology: Landscape and Protection*, Academic Press, London.

Jasper, R. G. (1999), 'The changing direction of land managers in reducing the threat from major bushfires on the urban interface of Sydney', in Lunt, I. and Green, D.G. (Eds.), *Bushfire 99: Proceedings of the Australian Bushfire Conference*, 7–9 July 1999, School of Environmental and Information Sciences, Charles Sturt University, Albury, NSW, pp. 175–184.

Jeffery, M. I. (1997), *Biodiversity and Conservation*, Routledge, London.

Kelly, A. H. and Farrier, D. (1996), 'Local government and biodiversity conservation in New South Wales', *Environmental and Planning Law Journal*, Vol. 13 No. 5, pp. 374–389.

Kelly, A. H. (2006), 'Urban amenity: does it coincide with biodiversity conservation at the local government level?', *Australasian Journal of Environmental Management*, Vol. 13 No. 4, pp. 243–253.

Kelly, A. H. and Stoianoff, N. P. (2006), 'Local government rates in New South Wales, Australia: an environmental tax?', in Cavaliere, A., Ashiabor, H., Deketelaere, K., Kreiser, K. and Milne, J. (Eds.), *Critical Issues in Environmental Taxation III*, Richmond Law & Tax Ltd, Richmond, UK, pp. 533–553.

Kelly, A. H. and Smith, C. (2008), 'The capriciousness of implementing Australian planning law: zoning objectives in NSW as a case study', *Urban Policy and Research*, Vol. 26 No. 1, pp. 83–100.

Leonard, J. (2003), 'People and property: a researcher's perspective', in Carey, G., Linenmayer, D. and Dovers, S. (Eds.), *Australia Burning: Fire Ecology, Policy and Management Issues*, CSIRO Publishing, Collingwood, Victoria, pp. 103–112.

Liverpool City Council and Ecological Australia (2003), *Liverpool City Council Biodiversity Strategy*, Liverpool City Council, Liverpool, NSW.

Little, S. J. (2002), 'New bushfire protection mandates for New South Wales', *Local Government Law Journal*, Vol. 8 No. 2, pp. 57–69.

Little, S. (2003), 'Preventative measures for bushfire protection', *Australian Planner*, Vol. 40 No. 4, pp. 29–33.

Maddocks, C. (2009), 'Sowing seeds of hope', *Spectrum, Sydney Morning Herald*, 18 July, p. 25.

Martin, P. (2009), 'Australia trumping even the US when it comes to McMansions', *Sydney Morning Herald*, 30 November, p. 1.

McAneney, J., Chen, K. and Pitman A. (2009), '100-years of Australian bushfire property losses: is the risk significant and is it increasing?', *Journal of Environmental Management*, Vol. 90 No. 8, pp. 2819–2822.

McAuslan, P. (1980), *The Ideologies of Planning Law*, Pergamon Press, Oxford.

McKinney, M. L. (2006), 'Urbanization as a major cause of biotic homogenization', *Biological Conservation*, Vol. 127 No. 3, pp. 247–260.

New, T. (2000), *Conservation Biology*, Oxford University Press, Melbourne.

NSW Department of Planning (2008), *NSW Housing Code: Guide to Complying Development for Detached Housing*, Department of Planning, Sydney.

NSW National Parks and Wildlife Service (2004), *Endangered Local Community Information: Cumberland Plain Woodland*, NSW National Parks and Wildlife Service, Hurstville, NSW.

NSW Rural Fire Service and PlanningNSW (2001), *Planning for Bushfire Protection: A Guide for Councils, Planners, Fire Authorities, Developers and Home Owners*, New South Wales Rural Fire Service, Granville, NSW.

NSW Rural Fire Service (2006), *Planning for Bush Fire Protection: A Guide for Councils, Planners and Developers*, New South Wales Rural Fire Service, Granville, NSW.

Odger, B., Ryan, M. and Wells, K. (2003), 'Forest, trees, people and fire: lessons for Canberra', *Australian Planner*, Vol. 40 No. 3, pp. 37–38.

Possingham, H. (2008), 'Biodiversity', in Lindenmayer, D., Dovers, S., Olson, M. and Morton, M. (Eds.), *Ten Commitments: Reshaping the Lucky Country's Environment*, CSIRO Publishing, Canberra, ACT, pp. 155–162.

Ramsay, C. and Rudolph, L. (2003), *Landscape and Building Design for Bushfire Areas*, CSIRO Publishing, Collingwood, Victoria.

Ramsay, G. C., McArthur, N. A. and Rudolph, L. (1995), 'Towards an integrated model for designing for building survival in bushfires', *CALMScience Supplement*, 4, pp. 101–108.

Riddell, G. (2005), 'A crumbling wall: the Threatened Species Conservation Act 10 years on', *Environmental and Planning Law Journal*, Vol. 22 No. 6, pp. 446–458.

Roddewig, R. (1978), *Green Bans: Birth of Australian Environmental Politics*, Hale & Ironmonger, Sydney.

Russell-Smith, J., Yates, C. P., Whitehead, P. J., Smith, R., Craig, R., Allan, G. E., Thackway, R., Frakes, I., Cridland, S., Meyer, M. C. P. and Gill, A. M. (2007), 'Bushfires "down under": patterns and implications of contemporary Australian landscape burning', *International Journal of Wildland Fire*, Vol.16 No. 4, pp. 361–377.

Smith, D. (1974), *Amenity and Urban Planning*, Crosby Lockwood Staples, London.

Stein, L. A. (2008), *Principles of Planning Law*, Oxford University Press, Melbourne.

Syme, G. J., Fenton, M. R. and Coakes, S. (2001), 'Lot size, garden satisfaction and local park and wetland visitation', *Landscape and Urban Planning*, Vol. 56 No. 3/4, pp. 161–170.

Taylor, K. (1999), 'Colonial picturesque: an antipodean Claude Glass', in Hambler, A. (Ed.), *Visions of Future Landscapes*, *Proceedings of 1999 Australian Academy of Science Fenner Conference on the Environment*, Bureau of Rural Sciences, Canberra, ACT, pp. 58–66.

Tuan, Y. (1974), *Topophilia: A Study of Environmental Perception, Attitudes and Values*, Prentice Hall, Englewood Cliffs, New Jersey.

Victoria Bushfire Reconstruction and Recovery Authority (2009), 'About the Victorian Bushfire Reconstruction and Recovery Authority', available at: http://www.wewillrebuild.vic.gov.au/about-us.html (accessed 18 November 2009).

Wilcox, M. R. (1967), *The Law of Land Development in New South Wales*, Law Book Company Ltd, Sydney.

Zagorski, T., Kirkpatrick, J. B. and Stratford, E. (2004), 'Gardens and the bush: gardeners' attitudes, garden types and invasives', *Australian Geographical Studies*, Vol. 42 No. 2, pp. 207–220.

10 China building control on green buildings

Rui Guan Michael

10.1 Background

10.1.1 Zoning concept in Chinese building control

China has a huge land mass covering southern tropical areas to northern cold fields. Based on climatic conditions, five zones are identified which further divide the country into seven climatic regions. Building regulations that relate to environmental design and energy conservation issues are designed accordingly.

I.   Severe cold zone
II.   Cold zone
III. Hot summer and cold winter zone
IV. Hot summer and warm winter zone
V.   Temperate zone

As shown in Figure 10.1, more than half of the country is located in the cold or severe cold zones. It further demonstrates that energy conservation is a key concern in many local building codes (CURAPD-GB 50180-1993, 2002).

10.1.2 Green buildings

Historically, energy efficiency control, as a key issue in many existing building regulations, covered the aspects of heating, ventilation, air quality, sun lighting, and building external envelope insulation. However, many of these regulations were written with one way of thinking – to secure the artificial built environment quality, but with very little concern for the con-sumption and support of natural resources. Recently, rapid urban growth with a high level of construction activity posed a heavy load on nature, and with more and more serious environmental problems appearing and stronger resources and energy worries arising, sustainable development became critical. Conse-quently, China's building control authorities established a few building

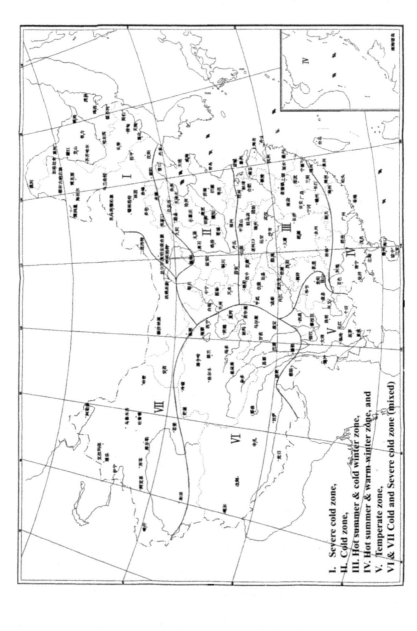

I. Severe cold zone,
II. Cold zone,
III. Hot summer & cold winter zone,
IV. Hot summer & warm winter zone, and
V. Temperate zone,
VI & VII Cold and Severe cold zone (mixed)

Figure 10.1 China climatic zoning map (for building and control)

Source: Adapted from *Code of Urban Residential Area Planning & Design, GB 50180-1993 (2002 version)*.

regulations for sustainable development and green buildings. Compared to earlier building codes, the new green building regulations have more concern for and protection of natural resources. They emphasize both built environment quality and sustainable support of nature and resources capability.

The latest Chinese green building regulation defines green building[1] with two layers of requirements. First, the building must fulfill peoples' functional requirements, and provide healthy, comfortable, and efficient usable spaces. Second, the building must economize resources such as energy, land, water and materials, and bring less pollution to the environment in its whole building life cycle. The key spirit of green building is the harmony and co-prosperity created between human development and nature.

10.2 Green building control

China's building control system has two major levels: the national and industrial level and the regional and corporate level. The higher level mainly constitutes building laws through the National People's Congress (NPC) and building regulations through the Ministry of Housing and Urban–Rural Development (MHURD), while the lower level publishes regional building regulations, building codes, and standards through local legislation and institutions.

Building regulations on both levels can be classified as compulsory and recommended regulations. The compulsory ones are comparatively more performance based. They are to secure basic issues of building quality, safety, health, and environmental concerns. Strict enforcement is required and legal punishment will ensue in the case of failures. On the other hand, the recommended regulations are more flexible. They could be adopted by local legislations or corporations. Many of them are prescriptive based with targeted issues or defined objectives.

Regulations are numbered as "GB, 50 Serial Number-Year" or "Industry Code Number, Serial Number-Year" for compulsory codes, and "GB/T, 50 Serial Number-Year" or "Industry Code Number/T, Serial Number-Year" for recommended codes. The following regulations include many earlier energy conservation and environmental control codes and the latest green building regulations.

- Standard for Sound Insulation Design of Civil Building Engineering, GBJ 118-1988
- Code of Urban Residential Area Planning & Design, GB 50180-1993
- Standard for Energy Efficiency Design of Civil Building Engineering, JGJ 26-1995
- Standard for Daylighting Design of Buildings, GB/T 50033-2001
- Code for Indoor Environmental Pollution Control of Civil Building Engineering, GB 50325-2001

- Design Standard for Energy Efficiency of Residential Buildings in Hot Summer and Cold Winter Zone, JGJ 134-2001
- Code for Design of Building Water Supply and Drainage, GB 50015-2003
- Codes for Design of Heating Ventilation and Air Conditioning, GB 50019-2003
- Design Standard for Energy Efficiency of Residential Buildings in Hot Summer and Warm Winter Zone, JGJ 75-2003
- Standard for Lighting Design of Buildings, GB 50034-2004
- Codes for Thermal Engineering Control of Civil Building, GB 50176-93
- Technical Code for Solar Water Heating System of Civil Buildings, GB 50364-2005
- Technical Specification for External Insulation on Walls, JGJ 144-2004, J 408-2005
- Design Standard for Energy Efficiency of Public Buildings, GB 50189-2005
- Code for Design of Outdoor Water Supply and Wastewater Engineering, GB 50013& 50014-2006
- Evaluation Standard for Green Building, GB/T 50378-2006

As shown above, many earlier regulations are compulsory, targeted more toward energy saving, heating, sun lighting, insulation, etc., but the latest Evaluation Standard for Green Building (ESGB) is a recommended regulation. A green award will be issued once a building is designed and built according to established standards. The three-star award is the highest national award, and a one- or two-star award represents a provincial or regional honor. Independent institutions are authorized to evaluate projects.

Six major aspects are identified in this green building regulation as key reviewed areas. Furthermore, every aspect has three levels of evaluation: the basic compulsory requirements, exclusive additional requirements, and classic additional requirements (ESGB-GB/T 50378, 2006).

- Land use and external environment design
- Energy saving and energy usage design
- Water saving and water resource usage design
- Material saving and material resource usage design
- Internal environment quality
- Building maintenance and management in its life cycle

To link this green building regulation to the previous established building codes, this new regulation takes many requirements from earlier codes as basic compulsory controls and prescribes higher standards for additional compliance with exclusive and classic controls. Any star-award green building must not only comply with all basic control requirements but also a certain percentage of the upgraded exclusive and classic control items shown in the chart below of "Green Building Regulation Controls and Awards in

Table 10.1 Green building regulation control and award in China

Green Building Regulation Control	Control Aspects	Total Control Items	Number (or Percentage) of Control Items Required for Green Building Award		
			One-star Award	Two-star Award	Three-star Award
Exclusive Control Items (total 40 items for domestic development and 43 items for civil public development)	Aspect 1: Land Planning & External Environmental Design	Domestic Development 8	4 50%	5 63%	6 75%
		Civil Public Development 6	3 50%	4 67%	5 83%
	Aspect 2: Energy Consumptions	Domestic Development 6	2 33%	3 50%	4 67%
		Civil Public Development 10	4 40%	6 60%	8 80%
	Aspect 3: Water Consumptions	Domestic Development 6	3 50%	4 67%	5 83%
		Civil Public Development 6	3 50%	4 67%	5 83%
	Aspect 4: Building Materials	Domestic Development 7	3 43%	4 57%	5 71%
		Civil Public Development 8	5 63%	6 75%	7 88%
	Aspect 5: Internal Environmental Quality	Domestic Development 6	2 33%	3 50%	4 67%
		Civil Public Development 6	3 50%	4 67%	5 83%
	Aspect 6: Management and Maintenance in Building Life Cycle	Domestic Development 7	4 57%	5 71%	6 86%
		Civil Public Development 7	4 57%	5 71%	6 86%
Classic Control Items		Domestic Development 9	–	3 33%	5 56%
		Civil Public Development 14	–	6 43%	10 71%

Source: Adapted from Evaluation Standard for Green Building, ESGB-GB/T 50378–2006.

China." This green building regulation currently controls two occupancy groups – domestic buildings and public civil buildings that only include office, commercial, and hotel developments. There are 76 requirements for the domestic group, including 27 basic control items, 40 exclusive items, and 9 classic items. There are 83 control items in total for the civil building group, including 26 basic items, 43 exclusive items, and 14 classic items. The following sections present green building regulation in the six aspects covering three levels of basic control and upgraded standards of exclusive and classic control.

10.2.1 Green building control on land planning and external environmental design

(A) Domestic development

There are eight basic control items for green residential development that cover the following aspects:

- Site selection
  The construction site shall not interfere with the local environmental balance. Farm land, natural water/river systems, forests, and cultural sites or historical regions shall be preserved.
- Land usage
  The land usage is controlled. For low-rise houses, the maximum average land use per dweller is 43 square meters, for medium-rise houses, it is 24 square meters, and for high-rise houses, it is 15 square meters.
- Internal environment
  Lighting, sunlight, and ventilation standards shall follow the existing code for Urban Residential Area Planning & Design, GB 50180-1993 (2002 version), which further regulates building envelope design, openings, and technical standards.
- Greenery
  Planting is controlled. Green ratio shall be not less than 30 percent and minimum 1 square meter per resident.
- Pollution release during construction process such as waste water, air, acoustics, etc., shall also be controlled.

Eight more items in exclusive control regulate issues including common facility design, reuse of old buildings, acoustic control detail, heat island index to be less than 1.5°C, external ventilation and wind control within residential community, planting selections and detail requirements, accessibility travel distance to public transportation within 500 meters, outdoor car park, green land and pedestrian design details including shadings and pavements, etc. The external rainwater dialyzable area including pavement, public green land and bare land shall be not less than 45 percent.

Two items in classic control emphasize the use of underground spaces and reuse of abundant sites and buildings.

(B) Civil public development

Five items in basic control similarly cover site selection, internal environment, pollution control, and environment protection during construction. At the exclusive control level, there are six items that regulate acoustic control standards, maximum external wind speed to be less than 5 m/s in outdoor pedestrian areas, roof design and vertical greenery, planting selection and design such as suggested mixed layers of arbor and shrubbery, accessibility to city transportation, and usage of underground spaces.

Three items in classic control focus more on reuse of abundant sites and old buildings, and external pavement design such as dialyzable area to be not less than 40 percent to release the heat island and deposit water content in soil.

10.2.2 Green building control on energy consumption

(A) Domestic development

Three basic control items regulate thermal design, heating system design, and air conditioning system design. Detail standards are referenced from the previous building code of Design Standard for Energy Efficiency of Public Buildings, GB 50189-2005. Six more exclusive control items focus on the following aspects:

- Site planning and building design, which covers control on building mass, facing, sun-shading, building block distance, and building envelope design like openness percentage for lighting and ventilation.
- Mechanical system design, which regulates the system selection and higher performance standards. For example, the centralized heating system and air conditioning system design should follow the established national code.
- Lighting design, which controls lighting selection in common areas in residential buildings. Sensitive lighting and high-efficiency lights are required.
- Design of energy reclaiming system when using central heating or air conditioning system in domestic buildings.
- Design and use of renewable energy such as solar energy and terrestrial heat. Such energy shall be more than 5 percent of overall building energy consumption.

Two classic items require higher standards of design of heating or air conditioning systems. The performance standard shall be less than 80 percent of the prescribed basic requirements and the renewable energy consumption ratio shall be more than 10 percent of overall energy consumption.

(B) Civil public development

According to national survey data, for civil public buildings such as mega commercial malls, hotels and office buildings, 50–60 percent of annual energy cost is consumed by heating or air conditioning systems and another 20–30 percent for lighting. In the heating and air conditioning, 20–50 percent is consumed through the building external envelope or shell. Therefore, China's building control imposes stronger regulations on energy aspects.

There are five basic control items regulating issues such the thermal design of a building: external envelope or shell, heating system and air conditioning system, prohibition of electronic water boiler heating systems, lighting design, and energy calculation methods.

Ten items in exclusive control detail standards as the following:

- Design of building layout shall consider local climate to incorporate with summer ventilation, winter sun-lighting, and winter strong-wind corridor prevention.
- Openness of external envelope control and openable window areas to be not less than 30 percent of overall external windows for natural ventilation.
- Controls on the window obturation design.
- Utilization of heat or cold storage techniques in building design and material selection.
- Reducing the energy load of heating/cooling fresh air through use of exhaust system.
- Subentry calculation method is suggested in all energy-cost systems such as heating, lighting, and air-conditioning systems to better control and improve building energy loads.

At the classic control level, four items are highlighted with higher control standards:

- Overall energy costs shall be not less than 80 percent of the prescribed requirements in the basic national standard.
- Combined cold, heat, and power generation systems to be used for distributed energy systems in buildings.
- Use of solar energy, terrestrial heat, and other renewable energy to provide heat, water, and electricity for buildings. Furthermore, such heat and water shall be more than 10 percent of the total amount of heat and water, and electricity shall also be more than 2 percent.
- Lighting controls shall fulfill GB50034-2004, Standard for Lighting Design of Buildings, which states detailed requirements on lighting selection, glare control, lighting power density, etc.

10.2.3 Green building control on water consumption

(A) Domestic development

Five basic control items cover water consumption issues. A water scheme is required at project planning and design stages to plan water resources and balance their use. The design of a piping system is regulated by detailed codes and landscape watering must use recycled water. The sanitation facilities are required to save at least 6 percent of normal water usage. Furthermore, recycled water shall also fulfill the relevant quality and cleanliness requirement.

Six exclusive control items regulate more aspects and higher standards. A plan for surface water and rainwater collection and use is required. All non-drinkable water including greenery water or car-wash water shall be resourced from recycled water or non-traditional water. High-efficiency landscape watering systems are also required. Pavements are to be designed for water penetration. Non-traditional water resources or recycled water consumption must make up at least 10 percent of overall water usage.

One classic control item imposes even higher standards of non-traditional water usage to be not less than 30 percent of overall water consumption.

(B) Civil public development

The basic and exclusive control items for green public buildings have similar requirements as for residential buildings. There are five basic and six exclusive control items covering aspects from water planning, water resources and usage, piping systems, landscape watering systems, sanitation design, etc. The exclusive control level lists more requirements on design of water meters. The usage of recycled water or non-traditional water shall be not less than 20 percent for office and commercial buildings and 15 percent for hotels. Furthermore, the classic control upgrades this minimum standard to 40 percent for office and commercial buildings and 25 percent for hotels.

10.2.4 Green building control on building materials

In this aspect, the green building regulation has similar control items for residential and civil public buildings.

At the basic control level, both groups have two items that regulate building internal materials and decoration design. Ten previous building codes are referenced as compulsory standards that cover aspects from material selection to construction details. Noxious emissions from various materials such as the limit of formaldehyde emission, radiotoxic control, etc., are highly controlled. These standards not only control the building elements such as floors, walls, and ceilings, but also the internal furniture, carpet, and other decoration materials.

In the exclusive control aspect, there are similar items for both groups:

- The building materials produced on site shall be not less than 70 percent of overall needed materials for residential development and 60 percent for civil public development. Furthermore, the production site shall be located within 500 meters of the construction site for both groups.
- The Cast-in-Situ Concrete shall be ready-mixed concrete for both groups.
- Building structure shall rationalize the use of high-performance concrete and high-strength steel.
- Categorization of building materials during the demolition process and reuse of renewable materials are required.
- Consideration should be given to the utilization of reusable materials at an early building design stage. Furthermore, the reusable materials shall be more than 10 percent in weight of overall building materials.
- With building performance secured, the usage of recycled man-made building materials is suggested, and the total quantum of its usage shall be not less than 30 percent of the kin materials.

For office and commercial buildings, open design and flexible internal partitions are recommended to reduce material waste in re-decorations.

Two of the criteria in classic control are the same for residential and civil public buildings. One is the regulation of building structure systems to have less environmental effect and material cost. The other is ensuring that the utilization ratio of recycled building materials is more than 5 percent.

10.2.5 Green building control on internal environmental quality

(A) Domestic development

Five basic control items regulate the internal environment of residential buildings in aspects of sun-lighting, acoustic control, ventilation, and air quality.

- For every domestic unit, there must be at least one living space/room that fulfills the requirement of sun-lighting standards regulated in the previous building code. For a unit with no less than four living spaces/ rooms, two space/rooms have to fulfill such requirements. The detailed control standards in different climatic zones are listed in the Table 10.2.
- Space/rooms for living, study, and kitchen shall have external windows and the window design shall follow Standard for Daylighting Design of Buildings, GB/T 50033-2001. The minimum value of the daylight factor (Cmin (%)) is 1 for living room, bedroom, study, kitchen, and 0.5 for dining room, washroom, and staircases. The critical bottom line of interior daylight illuminance is 50 lx for kitchen and living spaces, and 25 lx for the rest.

Table 10.2 Sunlight calculation standard for residential buildings in China

China Climate Zoning for Housing	I, II, III and VII Climate Zone		IV Climate Zone		V and VI Climate Zone
	Big Cities	Medium and Small Cities	Big Cities	Medium and Small Cities	
Standard Calendar Date for Sunlight Calculation	Severe Cold Day				Midwinter Day
Sunlight in Hours	≥2 hours	≥3 hours			≥1 hours
Calculation Period during A Day	08:00 to 16:00				09:00 to 15:00
Sunlight Calculation Point	Windowsill of the Lowest Habitational Unit*				

The most critical point for such sunlight calculation is required as 0.9 meter above the internal floor of the lowest habtational unit

Source: Adapted from Residential Area Planning & Design, GB 50180–1993 (2002 version).

- Building envelope is regulated in design and material selections for acoustic control. Performance standard is a maximum of 45dB (A) for living and bedroom daytime, and 35 dB (A) at night. Other detailed standards for doors, windows, and slabs are also imposed.
- Natural ventilation is required for every living space/room. The openness percentage of building external envelope for ventilation purpose shall be not less than 8 percent of the floor area for the hot summer and warm winter zone, and 5 percent for other climatic zones.
- Air pollutant concentrations such as free formaldehyde, benzene, ammonia, and TOVC are controlled according to details in the code for Indoor Environmental Pollution Control of Civil Building Engineering, GB50325-2001.

At the exclusive control level, there are six items imposing more requirements on the internal built quality of residential buildings.

- The field of vision for every domestic unit is to be considered and cross view shall be avoided. At least one washroom shall have an external window for a unit with more than two washrooms.

- No dew on the internal surface of floor, ceiling, and external walls or windows.
- The maximum temperature measured at roof and internal surface of east and west external walls in natural ventilated conditions shall fulfill the detailed requirements in codes for Thermal Engineering Control of Civil Building, GB 50176-93.
- An adjustable sunlight shading system and air quality supervision system are advised.
- When using internal heating systems or air conditioning systems, regulatable user end-units are suggested.
  One classic control item further suggests the usage of high-performance heat or cold storage materials for living and bedroom design.

(B) Civil public development

Six items in basic control regulate internal built quality of office, hotel, and commercial buildings.

- Most civil public buildings are using central air conditioning systems, so the internal temperature, humidity, and fresh air must follow the standards listed in Design Standard for Energy Efficiency of Public Buildings, GB 50189-2005.
- No dew at internal surface of building envelope.
- Air pollutant concentration such as free formaldehyde, benzene, ammonia, and TOVC is controlled according to relevant national regulations.
- Acoustic control is highlighted and design standards are referenced from the national code of Standard for Sound Insulation Design of Civil Building Engineering, GBJ 118-1988. For example, for offices and hotels, the minimum indoor background sound insulation is 45 dB.
- Building internal lighting, unified glare rating (UGR), general colour rendering index (Ra), and other design parameters are regulated to follow the national code of Standard for Lighting Design of Buildings, GB 50034-2004. For example, in office buildings, the minimum standard illuminance is regulated to be 300 lx, UGR 19, and Ra 80.

Six exclusive items further regulate higher control standards:

- Natural ventilation shall be well designed or applied to building design and construction detail design.
- Adjustable end-unit of air conditioning system is recommended.
- Sound insulation of building partition materials shall follow the first class of control in the national code of GBJ 118-1988.
- Building layout design shall consider zoning differences and external disturbance in acoustic control.

- More than 75 percent of all major functional space/rooms in hotels and offices shall fulfill requirements in the national code of Standard for Lighting Design of Buildings, GB 50034-2004.
- Design for accessibility shall be considered in building entries and major space/rooms.

Three classic control items suggest installing external adjustable sunlight shadings and air quality supervision systems to improve indoor thermal comfort and utilizing natural light to improve basement lighting.

10.2.6 Green building control on management and maintenance in building life cycle

(A) Domestic development

Building maintenance and management are important parts in the concept of the building life cycle of green buildings. At the basic control level, four items regulate residential buildings.

- The management company must provide a report of all resource consumption and energy savings. It shall include the usage of energy, water, materials, and other daily consumption. All resources shall be classified and used accordingly. The company also needs to submit a greenery management plan that covers the issues of rainwater collection, landscape watering, non-drinkable water resource usage, etc. The cost calculation is required to be classified by categories and dwelling units.
- Categorization, collection and transportation plan of rubbish and other domestic waste such as abandoned furniture must also be provided. The architectural design of the collection center must prevent pollution to the environment.

More requirements are regulated in seven exclusive control items.

- The design of the rubbish center should provide hydrant and drainage. Pollution and odor control should be considered in design.
- A high-tech supervisory controlling system is advised to save management manpower.
- Greenery must be maintained in good condition and the survival rate of planting and transplanting must be more than 90 percent.
- The management company must pass the ISO 14001 environmental management standard.
- Rubbish and waste is to be classified in collection and implemented by more than 90 percent of all dwelling units.
- Building maintenance facilities and the building piping system are to be designed for easy access of inspection and maintenance work.

There is one classic control item that regulates the issue of biodegradable rubbish. The rubbish shall be collected and kept separately. Pollution needs to be specifically controlled and effective methods such as washing and drainage must be considered.

(B) Civil public development

There are three basic control items that regulate the building maintenance and management work of civil public developments.

- A consumption savings report on energy and water, and a greenery management plan are needed.
- No polluted water or air is allowed to be released in the building lifecycle.
- A plan of categorization, collection, and transportation of all rubbish and domestic waste also must be provided.

Seven exclusive items set up more detailed requirements for building construction, mechanical, and electrical system design. The earthwork balance must be considered in a building construction plan. Roadways and other possible establishments during the construction stage are recommended to be utilized after building completion. Building ventilation systems shall be designed according to relevant established national codes and the higher performance standard. A high-tech supervisory controlling system is needed and the management company must pass the ISO 14001 environmental management standard. The previous energy cost is charged by area of square meter, which may lead to waste. Therefore, a new way of calculating electricity and heating/cooling energy costs based on unit consumption amount is suggested.

One classic control item encourages a reward system proposed by building management that links energy savings and operation costs to every building user.

10.3 Green building research and applied cases

10.3.1 Green building principle

In the coming years, China will have two billion square meters of construction area annually. More than 30 percent of national energy costs will be put into the construction and everyday usage of the buildings. Furthermore, building material fabrication will cost another 17 percent (Xiangjuan Kong *et al.*, 2008). Both items add up to nearly half of the national energy costs every year. The development condition is critical and green building is the key to sustainable growth. Four major principles are set up for green buildings in research and design aspects:

- Principle of Adjusting Measures to Local Conditions
- Principle of Life Cycle Assessment (LCA)
- Principle of Trade-off and Holistic Control
- Principle of Process Control

This huge country with diverse climatic zones requires localized controls. Principle 1 makes adaptability a key concern and allows flexibilities in building regulation control. The life cycle assessment (LCA) principle emphasizes the overall building impact on nature during the whole process. Four stages are considered in this life cycle concept. The first stage refers to all resource collection, production, and transportation including building materials, water, electricity, and other energy resources. The second stage focuses on planning, design, and construction. The third stage covers building operation, performance, and maintenance, and the last stage focuses on building demolition and issues afterward such as recycling materials and other resources. In Stages I and II, research and design concentrate on economizing resources and providing peoples' functional needs with healthy, comfortable, and efficient usable spaces. In Stage III, the focused aspect is long-term effectiveness of building usage, which involves issues such as daily energy cost, resource consumption, etc. The last stage is actually an evaluation of a building's impact on nature, particularly after its usage is complete. Through these stages, it hopes to provide guidelines for green building research and design and to achieve harmony between human development and nature. The "trade-off" principle is an important strategy in code making. It allows flexibilities in compliance with code requirements based on performance assessment. Such a principle works well with a holistic understanding of the core issues. The holistic control concentrates on overall effect and allows partial adjustment, minor conflicts, or interest sacrifice to achieve overall benefit. The "process control" principle regulates all procedures of building design, construction, operation, maintenance, and demolition. Compared to the second principle of LCA, which focuses more on technical issues, this principle focuses on the management aspect such as how concepts and designs could be correctly translated into built products and how those green buildings could be operated properly.

10.3.2 Green Olympic building assessment system

China's Green Olympic Building Assessment System (GOBAS) was published in 2003. It covers two aspects – an establishment of green Olympic building standards based on the green building concept and Olympic building requirements, and an operational assessment method that secures the higher standards of Olympic buildings through design and construction. As a key building control method, the assessment system would be introduced in the following sections:

(A) Assessment system

As shown below, the assessment process has four levels coherent with the overall construction processes.

- building planning
- building design
- building construction and quality acceptance
- building operation

Based on the requirements of each stage, different issues are focused upon in the assessment system. Building environment, energy, water resource, materials, and internal environmental quality are all well examined during the process. Passing an earlier stage is a compulsory step to enter the next stage of assessment (Yi Jiang *et al.*, 2004).

(B) Quality-load grading method

There are two contradictory sides for all green buildings. The positive side is that building construction should produce a good built environment for people internally and externally. The negative side is that such a construction process, its building product use, and even its demolition could be expensive in terms of resources and energy. Therefore, the whole building life cycle can impose a heavy load on the environment. The key task of green building is to balance minimizing resources, energy, and environmental load on one hand and maximizing built environment quality on the other. With such concepts considered, a balanced assessment method was developed for a holistic evaluation.

The Quality-Load grading method is a good way to balance both sides and present the "green" nature of the buildings more objectively. The items examined are classified in two groups as group quality (Q) and group load (L). Group Q refers to the artificial built environment quality and performance. Group L includes all necessary resources and energy cost. It refers to the environmental load that the building will impose on nature to achieve that guaranteed built quality in Group Q. For easier calculation, in Group L the load is valued as load reduction (Lr), which means the lighter the load a building causes on the environment, the higher score the building can get.

In Group Q, the aspects and items assessed are as follows:

Q1. Site Quality
- Disaster prevention control and emergency actions
- Atmosphere quality
Q2. Services and Functions Provided
- Civil infrastructure and facilities provision

- Public transportation provision
- Applicability of the building

Q3. External Physical Environment
- Environmental quality of noise
- Environmental quality of light
- Environmental quality of thermal comfort
- Environmental quality of wind
- Environmental quality of water
- Greenery

In Group Lr, five basic items and some subentries are:

Lr. 1. Necessity Study of the Project
- Necessity demonstration
- Project scale control
- Usage of temporary facilities

Lr. 2. Environmental Affect Analysis
- Land resource control
- Ecological environment control
- Physical environment control
- Civil infrastructure affect control

Lr. 3. Energy Consumption Analysis
- Energy planning
- Atmosphere affect analysis

Lr. 4. Materials and Resources
- Existing buildings
- Building materials
- Solid waste control

Lr. 5. Water Resource
- Water planning
- Waste/polluted water and rainwater control

Both groups are valued with five-point grades. Furthermore, based on the importance of each item valued, coefficients (weightings) are designed accordingly. So the green Q-L score of the assessed building is a summation of all products by each mark and its weight as shown in the following formula:

The Q-L Green Score = $\Sigma(5\ \text{point mark} \times \text{coefficient})$

For a mixed-use project or huge development with various components, the overall Q-L score also could be calculated based on floor area percentage of each part. In this way, a holistic description of the project's green nature could be quantified and further comparison with other projects could be made as well. As Figure 10.2 shows, there are five zones that represent the green nature of a building.

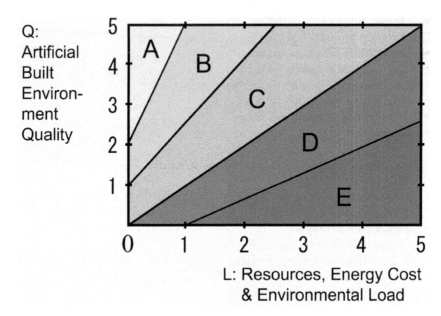

Figure 10.2 Quality–load, two-dimensional evaluation chart

Source: Adapted from *Green Olympic Building Assessment System Analysis*, 2004.

In Figure 10.2,

- Zone A represents a building that uses less resources and energy and has less environmental load but provides a better built quality. It is an excellent green building.
- Zone B represents a building that uses some resources and energy and has some environmental load to secure a good built quality. It is a good green building too.
- Zone C represents a building that uses more resources and energy and has more environmental load but provides good or general built quality. It is a green building, but with lower green quality.
- Zone D represents a building that uses a lot of resources and energy and has a stronger environmental load but provides general built quality. It is not a green building in nature.
- Zone E represents a building that uses or wastes even more resources and energy and has a very strong environmental load but provides lower built quality. It is not a green building in nature.

Buildings in Zones D and E are not green buildings. They have a strong environmental load and waste lots of resources. Particularly, buildings in Zone E should be avoided (Yi Jiang *et al.*, 2004).

The GOBAS could be used not only as an evaluation method but also as a basis for PLANNING and design guidelines. However, since the whole system is designed for assessment, the architects could only get evaluation data from the system as design reference. No compulsory building regulation controls could be followed. To enhance its broad application to the current building industry, further legislation work and integration with current building control systems are necessary. On the other hand, more market-based inspection tools could be developed for building designers. This national system could have localized versions with respect to provincial or climatic conditions.

(C) Comparison with other national systems

Many countries have developed their own assessment systems for green buildings, such as BREEAM (Building Research Establishment Environmental Assessment Method) in the UK, LEED (Leadership in Energy and Environmental Design) in the USA, CASBEE (Comprehensive Assessment System for Building Environmental Efficiency) in Japan, and NABERS (National Australian Building Environmental Rating System) in Australia, etc. The development of GOBAS makes good use of these systems and further localizes them within China's special context and links to existing building regulation control system. Like CASBEE, GOBAS uses the concept of double evaluation from positive and negative aspects in its grading method and formulates the Quality-Load grade chart. However, compared to CASBEE, GOBAS has very little subjective evaluation items in its planning aspect and focuses more on objective issues that are easily quantified (Zhiyong Zhang *et al.*, 2006). To control the general problem of over construction and improvident investment in China, the GOBAS sets up "Lr. 1 of Necessity Study of the Project" as a key variable to examine the issue. Project scale, construction quantum, plot ratio, and other variables are controlled in this aspect. In calculating the material usage ratio, the LEED system uses built area and material prices as a calculation base, which is easier for market applications, while GOBAS and CASBEE use material energy cost base, which includes more consideration of building material production, transportation, and demolition. Furthermore, to evaluate the energy cost of building materials in production, GOBAS brings its own calculation formula for standard evaluation. However, compared to LEED, a good market-based system with many simple design variables for users' easy understanding and manipulation, GOBAS still has a lot of improvements in its market promotions.

10.4 Summary and conclusion

China's building control mechanism was established in the 1960s. However, most building laws and building regulations were published in the past few decades. Many earlier regulations only focused on the quality of the artificial built environment and energy conservation issues, while the natural environmental load and support capacity were not considered much in code making. Only recently has green building research, its professional practice, and building control study been recognized.

In green building legislation aspects, China established its national control of Evaluation Standard for Green Building, GB T 50378-2006, in 2006. Currently, this regulation is not an absolute compulsory building code for practice. It has three control levels of basic, exclusive, and classic control level. At the basic level, many previously established building codes are referenced in this green building code as compulsory controls. Those regulated aspects focus more on the internal built environment and energy saving issues. The advanced requirements at the exclusive and classic control levels are not obligatory building regulations. Such conditions limit the sanction of green building control and further affect green design practice extension. To create a better and sustainable control system, both obligatory control and incentives are needed in building control strategy. On the other hand, building authorities could include more green building issues in the basic obligatory controls and maintain as strong encouragement in the recommended codes. For example, a faster channel of government authorization for qualified green design or building could help developers reduce their time cost and risk, which would further activate more participation in green building practices in China.

In green building authority control and management, to adapt to China's large land mass and local conditions, centralized national control could be coupled with provincial or regional controls as localization is one of the basic green building principles in China. Only fundamental laws should be made at the national level and more localized controls should be adapted to regional contexts. Such a code-making strategy could be adopted at both obligatory and promotional building control levels. It would also simplify green building control processes and invite more local participation.

In green building research and practice, GOBAS is a good assessment method that can be applied to more building types, used as design reference, and developed as green building codes. However, the contemporary practice of green building in China tends to be more high-tech. Many buildings that are labeled green are aggregations of different modern technologies and designed as landmark buildings. In fact, more research could focus on the general professional practice of normal green buildings and the study of these normal green buildings for common people and their lives. Many simple effective designs and techniques in vernacular architecture are neglected. In fact, green buildings need not strive for a "modern" landmark display if the traditional environmental wisdom in vernacular architecture or traditional Chinese buildings were truly understood, translated, and carried out.

Note

1   Green building is, according to the evaluation standard for green building (GB/T 50378-2006), a building that can economize resources such as energy, land, water, and materials, bring less pollution to the environment, fulfill people's functional requirements, and provide healthy, comfortable and efficient usable spaces in its whole life cycle.

References

Code of Urban Residential Area Planning & Design (CURAPD) GB 50180-1993 (2002) China Ministry of Housing and Urban-Rural Development (MHURD), Beijing, BJ.

Evaluation Standard for Green Building (ESGB) GB/T 50378-2006 (2006) China Ministry of Housing and Urban-Rural Development (MHURD), Beijing, BJ.

Jiang Y., Qin Y.G. and Zhu Y.X. (2004) "Green Olympic Building Assessment System Analysis," *Journal of China Housing Facilities*, Vol. 5 pp. 9–14.

Kong X.J. (2008) *Green Building and Low Energy Consumption Building – Cases Collection* (Lv Se Jian Zhu He Di Neng Hao Jian Zhu She Ji Shi Li Jing Xuan), China Architecture & Building Press, Beijing, BJ.

Zhang Z.Y. and Jiang Y. (2006) Assessment System for Green Building from the Angle of Ecological Design – Taking the CASBEE, LEED and GOBAS as Example, *Journal of Chongqing Jianzhu University*, Vol. 18 No.4, pp. 29–33.

11 Green buildings and the law in Taiwan

Jui-Ling Chen and Chiung-yu Chiu

11.1 Introduction

The development of green building requires a variety of effective and successive promotion efforts to accomplish an internalized transformation of industry and society. From the Earth Summit 1992 in Rio de Janeiro to the recent UN Climate Change Conference 2009 in Copenhagen, international summits not only frequently raised the international community's awareness of the crucial challenges of global climate change but also took actions to pursue sustainability during the past two decades. Following such a global pulse and confronting its own environmental predicaments, the government of Taiwan actively adopted concrete steps to come up with a series of policy instruments and implementation strategies for sustainable development. The concept of green building and its corresponding promotion programmes were thus initiated and have been incorporated into the national development plan as one of the most effective promotion measures in forging a better living environment. In order to address the topic of green buildings and the law, the chapter will be organized along the lines of the evolution of green building development in Taiwan. The entire evolution can be divided into three phases: technology research, policy promotion, and regulation implementation. The three ongoing stages can also form a feedback system so that the regulation of green building can meet the latest developments of green building technology. The "research" stage is typically aimed at identifying an adequate green building definition, to establish its assessment mechanism, and to develop relevant techniques. The "policy" stage is to promote the concept of green building, to facilitate the adoption of green building design with subsidies or incentives, and to conduct demonstrative projects that industry may follow. The "regulation" stage is to comprehensively implement green buildings through building regulations that industry must follow. Based on these three stages, the chapter is divided into three sections, green building research, green building policy, green buildings and the law, and followed by a concise conclusion.

11.2 Green building research in Taiwan

The green building technology research in Taiwan was devised from a series of systematic local surveys and studies. After many years of research, the Architecture and Building Research Institute (ABRI) of Taiwan proposed the ENVLOAD (Building ENVelope Thermal LOAD) as the earliest energy saving standard for building shell design in Taiwan and this was incorporated into Taiwan's building regulation in 1995. In order to further effectively alleviate the load of building construction and urban development to the environment with a broader aspect, the ABRI started to conduct a series of national green building research plans from July 1997, including the "Green Building and Living Environment Technology Plan," the first phase of which was from 1998 to 2001 and the second one from 2002 to 2006; "Green Building and Sustainable Environment Technology Plan" followed from 2007 to 2010; and there is a draft "Sustainable Green Building and Energy Saving Technology Plan," expected from 2011 to 2014. These plans and the expenses for each year are summarized in Table 11.1. The injection of more than 300 million NTD and the growing investment trend obviously

Table 11.1 Green building technology plans in Taiwan

No.	Title of technology plan	Year	Budget/ expenses*
1	Green Building and Living Environment Technology Plan (Phase I)	1998 1999 2000 2001	9,000 11,000 29,321 23,081
	Subtotal	72,402	
2	Green Building and Living Environment Technology Plan (Phase I)	2002 2003 2004 2005 2006	24,070 25,000 25,000 22,882 25,922
	Subtotal	122,874	
3	Green Building and Sustainable Environment Technology Plan	2007 2008 2009 2010	28,000 25,744 25,408 29,021
	Subtotal	108,173	
4	Sustainable Green Building and Energy Saving Technology Plan (draft)	2011–2014	to be ratified

*Unit: 1,000 NTD. Exchange rate: 1 USD equals to 29.5 NTD.

Source: Adapted from ABRI, 2005.

show a strong domestic interest in the green building technology development of Taiwan over the last decade.

The plans emphasize major topics, such as ecological environment protection, building energy saving, CO_2 emission reduction, resources recycling
and reuse, construction waste pollution prevention, indoor environmental
quality, building material applications, as well as the overall green building
industry development. Prior to discussing Taiwan's green building research
and development, the climatic condition of Taiwan and its local environmental issues that can fundamentally shape the nature of Taiwan's green building
will be introduced. One of the major outcomes of these technology plans,
that is, the establishment of the green building assessment system of Taiwan
and its corresponding certification mechanism, will then be explained.

11.2.1 Climatic conditions of Taiwan

Taiwan is situated southeast of the Asian continent. The Tropic of Cancer
(23.5°N) runs across the middle section of the island and divides the island
into two basic climates, the tropical monsoon climate in the south and
subtropical monsoon climate in the north. The topography is also one of the
factors affecting the climatic conditions. The temperature in mountain areas
drops as the altitude increases. Mountains over 1,000 meters high constitute
about 31 percent of the area of Taiwan. According to the statistics, Taiwan's
annual average temperature is about 24.7°Celsius in the south, 22.6°Celsius
in the north, and around 10.8°Celsius in central mountain areas as shown
in Figure 11.1.

Second, Taiwan receives abundant annual rainfall. The annual precipitation reaches about 2,500 mm, which also causes high humidity in Taiwan,
the average relative humidity being about 78 percent. Most rainfall is
concentrated in the summer, particularly during the periods when typhoons
visit. Since high temperature and high humidity characterize the climate of
Taiwan, the green building of Taiwan should properly reflect such climatic
conditions so as to ensure effective building energy saving design and to meet
the actual need of indoor thermal comfort.

11.2.2 Local environmental issues

With respect to the local environmental issues that Taiwan is confronting,
energy is the most challenging issue of all. Taiwan is highly dependent on
imported energy, which is currently over 99 percent of energy use. Out of all
energy consumption sectors, the building industry accounts for 28.3 percent
of the total energy consumption (including 0.4 percent building construction, 9.4 percent material production, 0.5 percent construction transportation, 12 percent housing energy use, and 6 percent commercial use), shown
in Figure 11.2. Moreover, the microclimate in metropolitan areas of Taiwan
has been getting warmer and warmer because of high urbanization (over

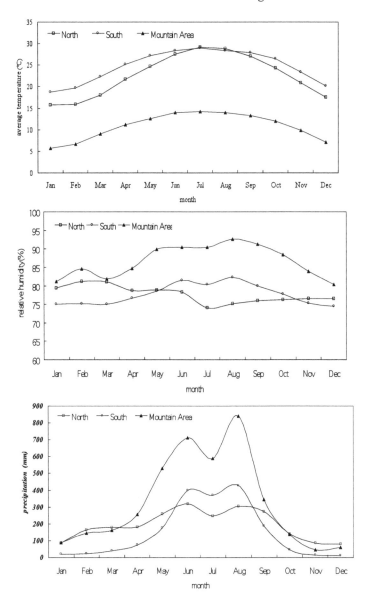

Figure 11.1 Average temperature, precipitation and relative humidity of Taiwan
Source: Adapted from the Climatic Statistical Data of Central Weather Bureau of Taiwan.

80 percent urban population), shortage of urban green spaces, imperme-
ability of public open spaces, and inefficient building energy saving design.
The higher temperature of the urban environment significantly aggravates
cooling energy consumption and CO_2 emissions, and accelerates the heat
island effect in the city. The temperature difference between downtown and

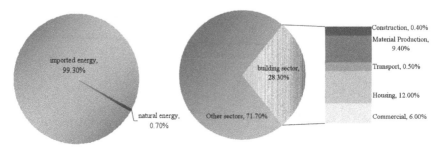

Figure 11.2 Current condition of energy consumption in Taiwan

the suburbs in most of the metropolitan areas in Taiwan is 3–4°Celsius in summer. According to the Taiwan Power Company's report, air conditioning electricity consumption increases about 6 percent when the outdoor air temperature increases 1°C. This indicates that the energy consumption of cooling in a building should be improved through better building energy-saving strategies, such as heat insulation, ventilation, and sun-shading design (ABRI, 2007a).

Despite the abundant rainfall and an average annual precipitation of more than 2,500 mm, the distribution of water resources is uneven, making the water available for use per capita relatively low. It is only one-sixth of the world's average, illustrated in Figure 11.3. Therefore, from the perspective of saving water resources, green building design and related promotion strategies in Taiwan should cover water conservation and reuse issues.

Another environmental issue emerges from the immense market of reinforced concrete structure, including steel and reinforced concrete (SRC), that comprises 86.80 percent of the entire building stocks in Taiwan, depicted as Figure 11.4. Reinforced concrete construction is typically considered as a high-polluting building method as it entails enormous energy consumption and CO_2 emissions, and uses natural river aggregate resources, as well as

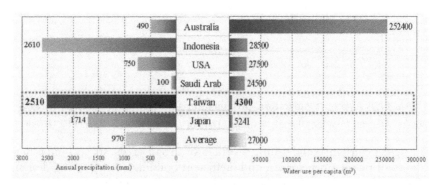

Figure 11.3 Comparisons of annual precipitation and water use per capita in Taiwan

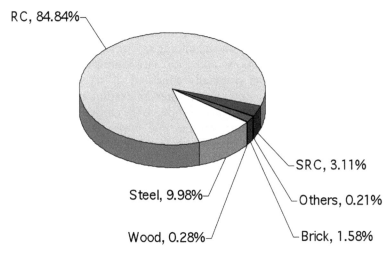

RC, 84.84%

SRC, 3.11%

Steel, 9.98%

Others, 0.21%

Wood, 0.28%

Brick, 1.58%

Figure 11.4 Distribution of 2009 building structure types in Taiwan
Source: Adapted from Construction and Planning Agency, 2008.

limestone exploitation. In Taiwan, the production of a ton of cement con-sumes 112 KWH, 134 kilograms of coal, and produces 450 kilograms of CO_2 emission (ABRI, 2008). Moreover, the lack of the mechanisms for recycling construction waste results in environmental pollution. A reinforced concrete building can generate 0.31 m³ of solid construction waste per square meter during the construction period and 1.23 m³ of waste per square meter in demolition (ABRI, 2008). Steel or wood structured buildings account for only a total of around 13 percent of buildings, which still need to be reinforced.

The fact that people stay indoors for about 90 percent of the time raises issues related to indoor air quality (IAQ), indoor environmental quality (IEQ), and indoor environmental health (IEH). These are being addressed and explored in depth. For example, with extensive material uses of interior finishing and remodelling for housing, the formaldehyde (HCHO) in build-ing materials and volatile organic compounds (VOCs) emitted in a warm environment can cause a high risk of people suffering from respiratory and skin diseases. From the perspective of sustainability and livability, a green building should be capable of providing a healthy living environment, for both its interior and exterior, with proper physical environmental design and quality building material application. Therefore, while various evaluation methods, assessment tools, and certification systems have been developed worldwide, the environmental performance assessment system for green building in Taiwan should be specialized and localized so as to accommodate the climatic characteristics (high humidity and high temperature), and correspond to its own local environmental issues, including energy saving, water conservation, waste reduction, as well as health.

11.2.3 EEWH green building assessment system of Taiwan

Introduced to address Taiwan's needs for a sustainable built environment described in previous sections, the green building assessment system capable of meeting the climatic conditions and local environmental issues was initiated in 1998. The system originally consisted of seven evaluation indicators: greenery (vegetation planting), water infiltration and retention, daily energy conservation, water conservation, CO_2 emission reduction, construction waste reduction, and sewage and waste disposal facility improvement. In 2003, due to the increasing interests in health and biodiversity issues globally, the ABRI modified the evaluation system, introducing two additional indicators: biodiversity and indoor environment quality. The current assessment system, with nine indicators, was thus finalized. These indicators can be further divided into four categories: ecology, energy saving, waste reduction, and health (now known as EEWH system), listed in Table 11.2. In addition to the assessment tool itself, a green building labeling system for green building certification was also established. The certification consists of two parts: Green Building Label for completed buildings, and Green Building Candidate Certificate for building projects. The minimum requirement for green building certification is to pass four indicators, including two prerequisites (daily energy conservation and water

Table 11.2 Evaluation indicators of green building assessment system of Taiwan

Category	Indicator	Evaluation items
Ecology	1. Biodiversity	Ecological net, biological habitat, plant diversity (only for sites greater than 1 hectare)
	2. Greenery	CO_2 absorption (kg-CO_2/(m². 40yr))
	3. Water infiltration and retention	Water content of the site
Energy Saving	4. Daily energy conservation (prerequisite)	Building envelope design ENVLOAD (20 percent higher than building regulation), and other techniques (including HVAC system, lighting, management system)
Waste Reduction	5. CO_2 Emission reduction	CO_2 emission of building materials (kg-CO_2/m²)
	6. Construction waste reduction	Waste of soil, construction, destruction, utilization of recycled materials
Health	7. Indoor environment	Acoustics, illumination, and ventilation, interior finishing building materials
	8. Water conservation (prerequisite)	Water usage (L/person), hygienic instrument with water saving
	9. Sewage and waste disposal facility improvement	Sewer plumbing, sanitary condition for garbage gathering

conservation) and two optional indicators from among the other seven. The label is valid for three years and renewable. The assessments of the nine indicators are operated independently in order to respond to the various impacts upon the earth environment, with corresponding quantitative calculation methods, equations, and criteria. Different from other major green building assessment tools developed by the private sector in other parts of the world, such as BREEAM and LEED, the EEWH green building assessment system of Taiwan is a government initiative and the certifying entity of the Green Building Label is the Ministry of the Interior of Taiwan, which has paved the way for further policy implementation and institutionalization.

The certification system remained as a pass/fail mechanism until 2006. According to a green building evaluation review in 2003, 85 percent of certified green building cases were simply flying over the minimum threshold of green building certification (passing four indicators) with another 9 percent of cases just doing a little bit better (passing five indicators), which represented a great potential for improving the green building design quality (ABRI, 2004). Through two years of research, a pilot version of a new green building rating system was launched in 2006 to aim at encouraging better green building design. Its scoring and classification algorithm was developed based on the database of previous green building evaluation results and existing assessment methods. Following a lognormal distribution (Figure 11.5), the rating system thus defined five classes of green building design, certified, bronze, silver, gold, and diamond, to encourage better green building practices and innovative design techniques.

The minimum requirement of four-indicator certification described above transferring into the scoring system is 12 points, scoring on a 100 point scale.

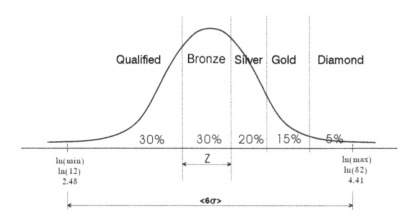

Figure 11.5 The lognormal distribution of previous green building evaluation results formed five classes of Taiwan's green buildings

Source: ABRI, 2007c.

The building or project receiving points from 12 to 26 can be classified as certified. The scores of a bronze-rated green building will be from 26 to 34, silver, from 34 to 42, gold, from 42 to 53, and diamond, greater than 53. The weighting factors of each indicator were obtained from an expert survey, listed in Table 11.3. The new rating system was officially executed in the beginning of 2007.

The plaque of Taiwan's green building label is shown in Figure 11.6. The rating system can also provide a basis for reference to devise further policy instruments with incentives and stimuli to encourage wider adoption of green building in the private sector. Since Taiwan started to promote the

Table 11.3 Weighting points for green building rating system

Indicator	Weighting points		
	Minimum	Maximum	Full score
1. Biodiversity	2 points	9 points	27 points
2. Greenery	2 points	9 points	
3. Water infiltration and retention	2 points	9 points	
4. Daily energy conservation			
(1) Envelope	2 points	12 points	28 points
(2) HVAC	2 points	10 points	
(3) Lighting	2 points	6 points	
5. CO_2 emission reduction	2 points	9 points	18 points
6. Construction waste reduction	2 points	9 points	
7. Indoor environment	2 points	12 points	27 points
8. Water conservation	2 points	9 points	
9. Sewage and waste disposal facility improvement	2 points	6 points	

Source: Adapted from ABRI 2007a.

Figure 11.6 Example of Taiwan's green building label

Source: ABRI, 2007c.

design of building energy saving and green building, the condition of high-energy consumption and environmental overload has gradually been eliminated. The subtropical and tropical building feature of Taiwan has also been characterized with deep-shading façades, building site layout with more green and permeable spaces, and natural landscape planning. By the end of 2009, a total of 2,418 buildings or projects were certified as green buildings with a total floor area of 31.24 million square meters, including 467 green building labels and 1,951 candidate certificates. Based on the latest statistics, since the rating system launched in 2007, there have been 14 diamond-rated green buildings, along with 26 gold-rated, 54 silver-rated, 153 bronze-rated, and 809 certified.[1] The percentage of green buildings just passing minimum requirements dropped to 76 percent. In comparison with the 2003 data, the result indicates significant progress in enhancing the green building design quality recently. Taipei Bei-tou Library was the first diamond-rated green building in Taiwan, presenting a subtropical feature of green architecture and a harmonized landscaping design with its surrounding natural environment (shown at the end of the chapter, Figure 11.17).

The assessment tool described above is only for new construction. The ABRI is currently extending the EEWH assessment system as a family, including EEWH-EC (Eco-Community) and EEWH-RN (for existing buildings), which is expected to cover larger-scale neighbourhood developments and the existing buildings, which account for 97 percent of total buildings in Taiwan.

11.2.4 Green building material evaluation and labelling system

In addition to green building design, building materials are one of the major components of the entire building industry. In order to achieve a comfortable and healthy living environment and to drive the building material industry upgrade, the ABRI established the Green Building Material (GBM) Evaluation and Labeling System in 2004, shown as Figure 11.7, based on a life cycle assessment concept. Its definition is an ecological, healthy, recycled, or high-performance building material that is capable of efficiently minimizing impacts to earth and environment and damage to human health during its entire life cycle (including resources exploitation, manufacturing, application, utilization, disposal, and recycling). The basic requirement of the GBM is to comply with national standards and relevant environmental protection regulations. For example, the prohibited substances should comply with heavy-metal-related test standards (see Table 11.4) and require no asbestos, no radioactivity, no toxic chemical substances, no inorganic halide or other chemical compounds according to the Montreal Protocol (CFC, halon, etc.).

The GBM system consists of four categories, including health, ecology, recycling, and high-performance (Ho *et al.*, 2008), depicted as Figure 11.8. The Healthy GBM for improving the indoor environmental quality requires low emissions on HCHO and VOCs and focuses on management and control following the ISO16000 test. The healthy GBM's standards include that

Figure 11.7 Green building material label of Taiwan
Source: ABRI, 2007b.

Table 11.4 Standards for heavy metal tests in Taiwan

Composition	Standards(mg/L)
Mercury(T-Hg)	0.005
Cadmium(Cd)	0.3
Lead(Pb)	0.3
Arsenic(As)	0.3
Haxavalent Chromium(Cr^{+6})	1.5
Silver(Ag)	0.05
Copper(Cu)	0.15

Source: Adapted from ABRI, 2007b.

HCHO is less than 0.08 mg/m^2·hr and TVOC less than 0.19 mg/m^2·hr. The ecological GBM typically includes low toxicity processing and natural materials without shortage crisis. The recycled GBM aims to reduce wastes and to reuse abandoned materials and recycled aggregates. As for high-performance, it refers to materials with high permeability, sound insulation, and energy saving. The high-performance GBM assessment includes ISO717-1, ISO717-2, ISO11654, and the test follows ISO140-3, ISO140-8, ISO354, and ISO9050, presenting the harmonization with the ISO standards in Taiwan's GBM system. The certifying entity has been elevated to the Ministry of the Interior and the label's valid period extended to three years from January 2010.

By the end of 2009, 292 GBM labels had been conferred, covering 2,966 green products. Among these products, healthy material occupies 78.77 percent, followed by high-performance, 13.7 percent, recycled 7.19 percent, and ecological 0.34 percent. The percentage distribution indicates that health issues

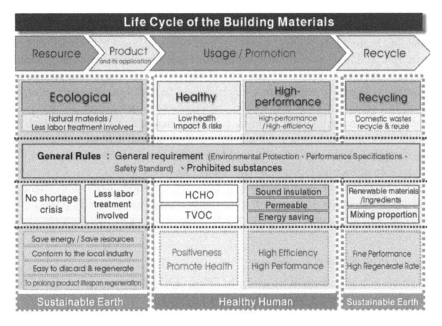

Figure 11.8 Framework of Taiwan's green building material evaluation system
Source: Ho *et al.*, 2008.

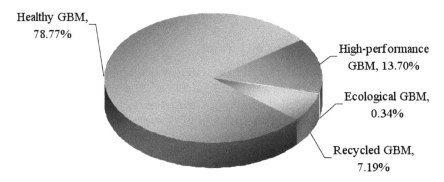

Figure 11.9 Percentage distributions of four green building material categories

have been highly emphasized and points to the development trend of the building material market in Taiwan, illustrated in Figure 11.9. The latest statistics
and relevant studies also show that the GBM labelling system is well coordinated with the current local green building evaluation practices. The GBM
system can also be applied to green building design and IEQ improvement.
Building materials that meet the standards of the GBM labelling system may
be used in various green building evaluation indicators. For example, indoor

building material construction evaluation and sound environment evaluation among indoor environment indicators encourage the use of all of healthy green building material, ecological green building material, recycled green building material, and high-performance soundproof green building materials. Specifically, preferential credits are adopted in the green building rating system to reward ecological green building material uses. In addition, the water infiltration and retention indicator would encourage the application of high-performance permeable green building pavement materials to possibly alleviate the impact of the urban heat island effect and reduce the load on public drainage facilities. The recycled green building materials meet the demand of CO_2 and waste reduction indicators that reduce the environmental burden caused by waste accumulation and new resource consumption. Regarding energy-saving glasses in the high-performance category, the materials may respond to the daily energy-saving indicator for efficiently reducing energy consumption.

11.2.5 Brief summary

Such an accomplishment of green building certifications or green building material applications cannot be achieved by simply creating green building assessment tools themselves without a complete set of effective promotion strategies. The green building policy actually played a key role in leading the development of green building in Taiwan, with both green building design and green building material application being institutionalized. The following section will introduce the governmental green building policy of Taiwan in detail.

11.3 Green building policy in Taiwan

Although the green building assessment system was adopted by the Ministry of the Interior of Taiwan as a national guideline for green building design and certification, it remained a voluntary mechanism until 2001 when the "Green Building Promotion Program (GBPP)" was ratified by the Executive Yuan[2] of Taiwan, which implementation period lasted from 2001 to 2007. Its implementation was a turning point in green building development in Taiwan, for starting the mandatory EEWH design in the public sector. Although its legal hierarchy was an administrative direction that only covered governmental buildings, green building practices in the public sector still successfully played a leading role in gradually forming a green building market. Furthermore, some indicators and their evaluation methodology extracted from the EEWH system for the newly constructed building design were then institutionalized in 2005. The green building policy of Taiwan has proved to be a globally unique initiative that has essentially forged a comprehensive promotion mechanism providing resources, regulations, research, guidance, training, and education to support the adoption of green building. Even after the completion of green building institutionalization work, the second-stage "Eco-City and Green Building Promotion Program (ECGBPP)" implemented from 2008 to 2011 continues its function in strengthening green building beyond the baseline set up by the law. The main contents of Taiwan's green building policy are summarized as follows.

11.3.1 Mandatory green building design for all new publicly owned buildings

The green building promotional strategy was to start with publicly owned buildings and to encourage private sectors to adopt the green building concept, so as to gradually evolve a common practice for the entire building industry. The mandatory EEWH green building design required publicly owned building projects with a construction cost greater than 50 million NTD to receive green building candidate certificate prior to the issuance of a building permit (as Figure 11.10), for all central governmental buildings since 2001 and for all local governmental buildings since 2003. Figure 11.11 clearly presents the dramatic effect of the mandatory green building design strategy. The increasing certified green buildings and projects showed a soaring trend after 2003. Such a mandatory requirement also showed a significant saving on electricity and water consumption. By the end of 2009, 460 buildings were certified as green buildings, and 1,908 projects received candidate certificates. The total accumulated electricity saving reached 788.8 million KWH, which was equivalent to 530.6 million CO_2-kg/year[3] and water saving reached approximately 34.89 million tons per year.

Due to the mandatory EEWH green building design requirement for the public sector, however, the statistics demonstrated an obvious fact that certified green buildings were concentrated on green building candidate certificates by 2007, where the green building label cases accounted for

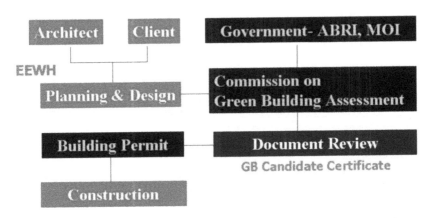

Figure 11.10 Conceptual framework of the mandatory EEWH green building design for all publicly owned buildings

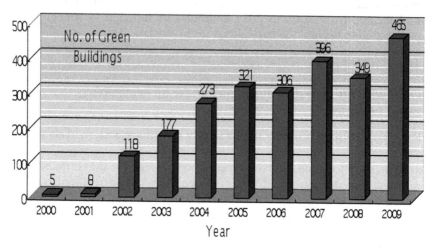

Figure 11.11 The growing trend of the numbers of Taiwan's certified green buildings in the last decade

17.46 percent.[4] Therefore, the second-stage promotion programme, "Eco-City and Green Building Promotion Program (ECGBPP)," proposed by the ABRI, started to strengthen the mandatory EEWH design for publicly owned buildings. The requirement was elevated to the acquisition of a green building candidate certificate prior to construction and, furthermore, a green building label prior to the official acceptance and auditing process (see Figure 11.12). Such a requirement aimed at ensuring the proper implementation of green building design, as well as controlling the quality of the buildings' environmental performance. After the implementation of the new controlling strategy, the green building label has had about 27 percent of

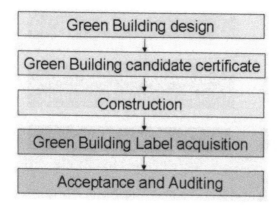

Figure 11.12 Strengthening mandatory EEWH green building design for the public sector

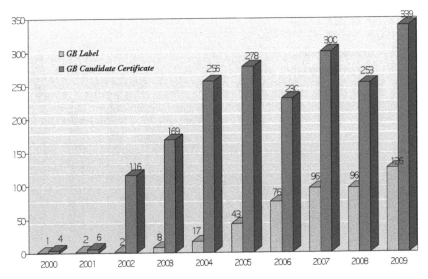

Figure 11.13 Statistics of green building label and candidate certificate from 2000 to 2009

total certified green buildings in the last two years and there is an increasing trend towards the future (see Figure 11.13).

A feature of the EEWH design requirement is that most of the certified buildings or projects are in the public sector.[5] The latest statistics demonstrate nearly 90 percent of green building certification in the public sector, depicted in Figure 11.14. The result, however, does not imply that most of the new buildings in the private sector are not green. Most of them are following the green building design standards in the Building Technical Regulations. Even so, such a wide gap does indicate great potential in further promoting the EEWH green building design in the private sector.

11.3.2 Green renovation projects for existing buildings

In Taiwan existing buildings account for about 97 percent. The most common problems of these buildings include over-capacity design of the HVAC system, deteriorating chiller efficiency, poor environmental quality, and ignorance of the ecological concept. If these buildings can be renovated in the following aspects, such as energy saving, water saving, water retention, and greening, the energy and resource consumption will be reduced and the ecological environment protected. Therefore, the ABRI has subsidized and conducted green renovation projects for governmental buildings since 2002 that focus on the issues of ecology, energy saving, waste reduction, and health for central government offices and national colleges and schools, while the Construction and Planning Agency of the Ministry of the Interior (CPAMI) concentrated on subsidies in the private sector. By the

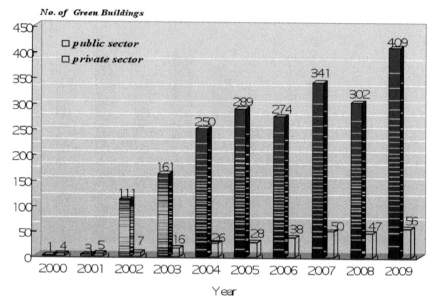

Figure 11.14 Statistics of green buildings in both public and private sectors from 2000 to 2009

end of 2009, eight major works had been undertaken, with the completion of 769 subsidized projects, listed in Table 11.5 including a green remodelling project for governmental buildings, a green HVAC project for governmental buildings, a building envelope heat insulation improvement project, an indoor environmental quality (IEQ) improvement project, greening existing buildings for the private sector, green diagnosis and renewal for existing governmental buildings, a building energy efficiency upgrade programme, and IEQ diagnosis and investigation projects. Among them, 456 projects were in the public sector and 313 in private. These projects and their outcomes are briefly summarized as follows.

Various green building remodelling techniques were adopted in these projects: (1) revitalization of the ecological environment; (2) improvement of water infiltration and retention; (3) employment of building energy-saving technologies on the building envelope and HVAC systems; (4) adoption of building energy-management strategies; (5) sustainable utilization of water resources; (6) mitigation of urban heat island effects; (7) reduction of light pollution; as well as (8) IEQ diagnosis and improvement work covering air quality, acoustic environment, thermal comfort, illumination, and electronic/magnetic impacts, illustrated as Figure 11.15.

Table 11.5 Recent green renovation projects for existing buildings in Taiwan

Projects	No. of cases	Year
1. Green remodeling project for governmental buildings	99	2002–2007
2. Green HVAC project for governmental buildings	97	2003–2007
3. Building envelope and roof heat insulation improvement project	336	2003–2007
4. Indoor environmental quality diagnosis and improvement project	18	2004–2007
5. Greening existing building demonstrative project for private sector	69	2004–2011
6. Green diagnosis and renewal for existing governmental buildings	42	2008–2011
7. Building energy efficiency upgrade program	66	2008–2011
8. Indoor environmental quality diagnosis and investigation project	42	2008–2011
Total	769	by the end of 2009

Source: Cheng, 2009.

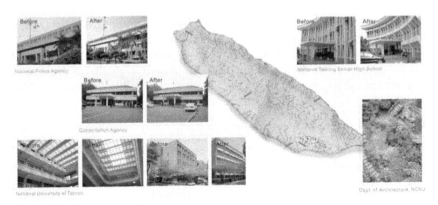

Figure 11.15 Diagram of the green remodeling projects in Taiwan
Source: ABRI, 2007.

11.3.3 Promotion, education, and international interchange

The ABRI also promoted the green building concept through a series of activities, including seminars, training courses, conferences, technical tours, green building award competition, and a green building expo. They encouraged outstanding architects to respond to the sustainable development and green building policy by adopting green building concepts in better design;

the ABRI has conducted the competition of the Green Building Award of Excellence annually since 2003. Through the competition, the award-winners were selected and recognized in public, so as to inspire more quality green building design and to expand people's identification of green building. By the end of 2007, 52 green buildings received the award of excellence. The competition provided substantial prizes (250,000 NTD) for each award winner. This successfully encouraged architects to learn more about green building and upgraded the design quality before the current rating system was implemented.

As for popularizing the green building concept and promulgating green building education, the ABRI has conducted a series of workshops, seminars, training courses, and technical tours in recent years and more than 10,147 delegates have attended. In order to promote the development experiences and facilitate the sharing of latest research in green building, the ABRI conducted green building forums in 2003 and 2004, and has regularly hosted the annual Subtropical Green Building International Conference from 2005 to 2007 as well as the 2009 Conference on Green Building – Towards Eco-City (co-sponsored by the CIB). This has allowed information on advanced research and innovative technology undertaken by many scholars and professionals from the countries in the subtropics and tropics to be transferred and for them to learn from each other. In 2005, the ABRI also assisted in forming the Taiwan Green Building Council as an exchange platform and joined the World Green Building Council. Through actively participating in major international events, such as SB04 in Kuala Lumpur, Malaysia, SB05 in Tokyo, Japan, SB07 in Taipei, Taiwan, GreenBuild07 in Chicago, USA, and SB08 in Melbourne, Australia, the green building evaluation tool and design concepts for the subtropics can be shared with the international community.

11.3.4 From green building to eco-city

The surge of the green tide worldwide represents an inevitable trend and requires a long-term commitment. Sustainable development has become an imperative for each country. The ABRI of Taiwan is currently implementing the ECGBPP to extend the scope of green building promotion into the community and city to show Taiwan's successive endeavors in global sustainability with local actions. The effort aims to develop eco-cities based on current green building accomplishments in a wide response to global climate change and to mitigate the heat island effect so as to ultimately achieve homeland sustainability. Finding the path from green building towards eco-community or eco-city does require more speculation. The ABRI first started by developing the evaluation system for eco-community (EEWH-EC) and reviewing the current regulations for urban planning and urban design. The system is expected to extract operable elements from diverse concepts in ecological, social, and economic aspects, and set up a systematic and quantitative evaluation method, so that planners and developers can follow clearly. The pilot version of the EEWH-EC was completed at the end of 2009. It involves green building certification credits in the system. Second, as with green building promotion experiences, the research results should be capable of being adopted in regulations, such as the Urban Planning Comprehensive Review Regulations and Urban Design Guidelines. A series of studies have been completed by the ABRI since 2008.

11.4 Green buildings and the law in Taiwan

As the ABRI is in charge of national architectural research in Taiwan, the Construction and Planning Agency, Ministry of the Interior (CPAMI) of Taiwan, is mainly in charge of legislative work on building and urban laws. Green building institutionalization was one of the major achievements of the GBPP. The CPAMI thus announced in 20 March 2004 the revision of the Building Technical Regulations, adding the Green Building Standard chapter that had been effective since 1 January 2005. Although the criteria of green building design in the building regulation is slightly lower than the EEWH system, the institutionalization work has now expanded compulsory green building design into the private sector.

11.4.1 Green building institutionalization

Following the mandatory regulation of green building design in the public sector, the green building design techniques and requirements, extracted from the EEWH system, have now been incorporated into Chapter 17 of the Building Technical Regulations (known as the Green Building Standard Chapter) by the CPAMI. The CPAMI is the designated agency to enact the Building Technical Regulations based on the Building Act.[6] The Green Building Standard Chapter was announced on 20 March 2004 and become effective from 1 January 2005. All newly constructed buildings must comply with green building design (see Table 11.6). The requirements cover general design guidelines and five specific categories of green building techniques, including greenery (vegetation planting), water infiltration and retention, building energy saving, rainwater and grey water reuse, as well as green building materials. Each specific category has its corresponding eligibility, summarized in Table 11.7, with an approximate 78 percent coverage of all newly constructed buildings currently.

Due to the subtropical and tropical climatic conditions of Taiwan, green building energy saving design divided Taiwan into three climatic zones: north, central, and south (§308 through §312). Energy-saving-related design and its corresponding criteria, such as heat penetration rate of roofs and exterior walls, windows, and openings, ENVLOAD, etc., follow this zoning model. The design standard in the Building Technical Regulations is slightly lower than that in the EEWH system in order to achieve a wider promotion

Table 11.6 Green building standard chapter in the Building Technical Regulations of Taiwan (effective since July 2009)

Items*	Articles	Effective date
General Design Guidelines	§298-301	January 1, 2005
Greenery (Vegetation Planting)	§302-304	January 1, 2005
Water Infiltration and Retention	§305-307	January 1, 2005
Building Energy Saving	§308-315	January 1, 2005
Rainwater and Grey Water Reuse	§316-319	January 1, 2009
Green Building Materials	§321-323	5% since July 2006 30% since July 2009

*§320 regarding the green structure has been cancelled since January 2009.

Source: Adapted from CPAMI 2009a.

Table 11.7 Eligibility of the green building design in the Building Technical Regulations of Taiwan

Items	Eligibility
Greenery	1. Schools 2. High-rise buildings more than 60 meters high or 16 floors 3. Buildings on hillside 4. A building site in the urban planning area for which the land use zoning code is residential area, educational area, scenic area, government agency, commercial area, or market, and conforms with the following requirements: (1) that is adjacent to the road with a width of more than 8 meters with a consecutive connection of more than 25 meters; or (2) that the total floor area is more than 1,000 square meters for commercial uses and market, or more than 1,500 square meters for residential area, educational area, and government agency.
Water infiltration and retention	Same as above except buildings on hillside.
Building energy saving	Schools, residential, large-space buildings (e.g. theater, stadium, etc.), and any other building with a total floor area of more than 1,000 square meters. Exemptions include machinery rooms, warehouses, cottages less than 500 meters, greenhouses, etc.
Rainwater and grey water reuse	A building with a total floor area of more than 30,000 square meters. Exemptions include industrial uses, warehouses, hospitals, etc.
Green building materials	Buildings for public uses.

and application. In July 2006, at least 5 percent of green building material utilization was adopted in the national building regulation as well. The percentage was further increased from 5 percent to 30 percent on 1 July 2009. Such a regulation provides a new opportunity for building material industry transformation and the CPAMI also devised a series of design guidelines for detailed calculation and evaluation for designers' use. The overall green building institutionalization was thus completed with all articles (§298 through §323) effective since 1 January 2009 after implementation in different phases.

11.4.2 Urban regulations

Review of urban regulations considering the eco-city concept is one of the core works of the current ECGBPP. The ECGBPP of Taiwan has only been implemented for two years and the link between green buildings and the urban-related regulations of Taiwan is currently under consideration. There are three regulations related to urban issues, listed as follows:

(1) Urban Planning Regular Comprehensive Review Regulations

The Regulations are enacted based on the Urban Plan Act §26. The urban plan shall be reviewed by the responsible agency of the government every three to five years. The review items include natural and human cultural resources, population and its density, building density, industrial structure and development, land use, infrastructure capacity, housing demand and supply, as well as transportation. The revision in response to eco-city issues on the Urban Planning Regular Comprehensive Review Regulations is currently undertaken by the Ministry of the Interior. The major revision is concentrated on parks and green spaces. Article 16 of the current regulations specifies the minimum area of parks depending on the neighbourhood population. In the future, the concept of "ecological network" is planned to be incorporated into the regulations so as to increase the connection of urban green spaces in addition to area requirements.

(2) Urban Design Guidelines

Unlike the Building Technical Regulations and Urban Planning Regular Comprehensive Review Regulations, the source of the Urban Design Guidelines is from the Urban Planning Regular Comprehensive Review Regulations §8. The eligible areas include new town, new urban development area, urban renewal area, historical and other preservation area, areas located within one kilometer two sides of high speed rail, highway, or specified scenic roads, as well as other areas designated on the master plan. The contents vary from local needs, including open space system configuration, pedestrian space and sidewalk systems, transportation system, detailed constraints on building site development, building details (height, styles, colours, etc.), environmental protection facilities, landscaping, as well as management and maintenance. In general, urban design is reviewed by commissions organized by local governments. The Urban Design Guidelines have been adopted by many local governments in Taiwan, whether large metropolises (Taipei and Kaohsiung) or regional cities (Taichung and Tainan). Among them, the Tainan municipal government set up its own general Green Building and Urban Design Guidelines, effective since 15 April 2004. The general guidelines follow the EEWH assessment system and have two prerequisites (daily energy conservation and water conservation) for city-owned buildings that cost between 20 and 50 million NTD. Other green building techniques such as vegetation planting, water infiltration design, and ecological engineering, have been included in individual urban design guidelines.

(3) Regulations of Urban Renewal Incentives on Floor Area

The Regulations of Urban Renewal Incentives on Floor Area were enacted based on the Urban Renewal Act §44. It was first put into practice in 1999. The latest revision was green-building-related and has been effective since 15 October 2008. The Regulations of Urban Renewal Incentives on Floor Area §8 allows the incentives on a maximum 10 percent extra development floor area for an EEWH silver-rated green building. The article caused many debates from opposite viewpoints during discussion. The urban renewal project requires much negotiation and processing time and the effects of such an article have not yet been investigated or reported.

11.4.3 Environmental impact assessment review directions

This paragraph is a complementary reference for green building and the laws in Taiwan. In addition to the laws in the building sector, the Environmental Impact Assessment (EIA) Act administered by the Environmental Protection Agency (EPA), Executive Yuan, contains a series of regulations related to green buildings. According to the EIA Act §21, eleven development activities where there is a concern of adverse impacts on the environment require an EIA to be conducted. The EIA is reviewed by a committee organized by the EPA following the organization rules. The concept of the EEWH green building design was incorporated into some of the EIA processes as one of the administrative directions for the committee effective since 2000. This was five years earlier than green building institutionalization. The EIA Review Directions for green building design cover four development activities, including for residential areas (§21; 14 September 2000), industrial areas (§34; 19 April 2000), golf courses (§26; 17 December 2001), and cultural, educational, and medical areas (§26; 17 December 2001).

11.5 Conclusions

In conclusion, all new buildings in the public and private sectors are required to have mandatory green building design, either based on the EEWH design or the Green Building Standard Chapter in the Building Technical Regulations. For old buildings, many green renovation projects for existing buildings and schools currently remain in progress. Significant savings on electricity and water resources can be expected to accumulate year after year. The research, policy, and regulation can also form a feedback system and bring out many opportunities for architects, professional consultant services, and the entire building and construction industry to assist in targeting a sustainable conversion. Today the market has been changed indeed. The market for water-efficient fixtures has risen by about 40 percent and the 12-litre toilet has vanished from the market. Globally leading semi-conductor manufacturing companies of Taiwan have started a large green building movement in the industrial sector. For example, in the Southern Science and Industry Park in Tainan County, Taiwan, 26 green buildings have been recently erected and among them three buildings are EEWH diamond-rated, as shown at Figure 11.16. This shows a demonstrative transformation for these companies in pursuing both industrial and environmental sustainability.

In addition, the outputs derived from policy implementation and law enforcement can actually feed back to inspire research topics. Such feedback may form a vital circle that is able to upgrade green building technologies and to gradually transfer the green building market from distinctive cases into the industry mainstream. Progress from design concepts into certification mechanisms or law provisions is not the final destination of green building promotion; the ABRI of Taiwan will keep an energetic momentum going to collaborate with the international community in conducting

Figure 11.16 Diamond-rated green building in the South Science and Industrial Park
Photographer: Wen-Hung hu, 2010.

Figure 11.17 Taipei Bei-tou Library, the first diamond-rated green building in Taiwan

Photographer: Wen-Hung hu, 2010.

advanced architectural research and viable policy promotion in order to achieve our better common future.

Notes

1 A total of 1,362 cases that have passed green building certification by 2006 are not applicable.
2 The Executive Yuan is the highest administrative organ of Taiwan.
3 CO_2 emission reduction obtained from only two prerequisites (electricity and water saving) are calculated.
4 The mandatory EEWH design requirement for the public sector also caused most of the cases just passing the minimum threshold before 2007.
5 Among the total of 2,418 green buildings, there are 2,141 cases in the public sector and 277 in private.
6 The Building Act §97 specifies the CPAMI to enact the Building Technical Regulations to regulate building planning, design, construction, structure, and equipment (CPAMI, 2009b).

Bibliography

Architecture and Building Research Institute, MOI, *Introduction to Green Remodeling Projects for Governmental Buildings*, 2004.
Architecture and Building Research Institute, MOI, *Tenth Anniversary Review of the ABRI*, 2005. (Chinese)

Architecture and Building Research Institute, MOI, *Good to be Green*, 2006.

Architecture and Building Research Institute, MOI, *Evaluation Manual for Green Buildings in Taiwan, 2007 New Edition*, 2007a. (Chinese)

Architecture and Building Research Institute, MOI, *Evaluation Manual for Green Building Materials, 2007 New Edition*, 2007b. (Chinese)

Architecture and Building Research Institute, MOI, *Good to be Green Brochure*, 2007c.

Architecture and Building Research Institute, MOI, *Challenges for the Subtropical Green Building*, 2008. (Chinese)

Construction and Planning Agency, MOI, *Statistics of Building Structure Types*, 2007 http://www.cpami.gov.tw/web/index.php?option=com_content&task=view&id=948&Itemid=75 (accessed on 17 June, 2008). (Chinese)

Construction and Planning Agency, MOI, *Building Technical Regulations*, 2009a. (Chinese)

Construction and Planning Agency, MOI, *Collection of Building Laws*, 2009b. (Chinese)

Construction and Planning Agency, MOI, *The Statistical Yearbook of Construction and Planning Agency Ministry of the Interior of 2009, Construction Licenses – By Materials*, http://w3.cpami.gov.tw/statisty/98/98_pdf/06_building/36120.pdf (accessed 19 December 2010). (Chinese)

Executive Yuan, *Eco-City and Green Building Promotion Program*, 2008. (Chinese)

Environmental Protection Agency, Executive Yuan, *Environmental Impact Assessment Act*, http://law.epa.gov.tw/en/laws/379692190.html (accessed 31 October 2009).

Environmental Protection Agency, Executive Yuan, *Environmental Impact Assessment Administrative Directions*, http://ivy5.epa.gov.tw/epalaw/search/LnameTypeList.aspx?ltype=03&lkind=3 (accessed 31 October 2009). (Chinese)

Chen, Jui-Ling and Chiu, Chiung-Yu, "Green Building Policies in Taiwan," *Proceedings of the Conference on SB04 South East Asia*, Kuala Lumpur, Malaysia, April, 2005.

Cheng, Yuan-Liang, "Green Renovation for Existing Buildings in Taiwan," *Proceedings of the 2009 Conference on Green Building – Towards Eco-City*, Taipei, Taiwan, October 2009.

Ho, Ming-Chin and Chiu, Chiung-Yu, "Introduction to Green Building Policy in Taiwan," *CIB W62 Symposium 2006*, Taipei, Taiwan, September 2006.

Ho, Ming-Chin, "A Sustainable Conversion in Building Industry: Green Building Promotion Program in Taiwan," *Proceedings of the 4th CECAR*, Taipei, Taiwan, June 2007.

Ho, Ming-Chin, *et al.*, "Taiwan Green Building Material Labeling System and its Applications to Sustainable Building in Subtropical Zone," *Proceedings of SB World Conference*, Melbourne, Australia, September 2008, 435–440.

Ho, Ming-Chin, "From Green Building to Eco-City," *Proceedings of the 2009 Conference on Green Building – Towards Eco-City*, Taipei, Taiwan, October 2009.

12 Climate change and the construction industry: sustainability challenges for Singapore

Asanga Gunawansa

12.1 Introduction

There is overwhelming scientific consensus that since pre-industrial times increasing emissions of greenhouse gases (GHGs)[1] have led to a marked increase in atmospheric GHG concentrations (IPCC, 2007), causing global warming. The causes for this phenomenon, more popularly known as climate change, can be divided into two categories, namely, natural causes and those that are created by man. The natural factors responsible for climate change include the continental drift, volcanoes, ocean currents, the earth's tilt, and comets and meteorites. The human factors include industrial wastage, use of natural resources for human consumption and habitation, population increase and the extreme reliance on the use of fossil fuels and natural gases.

Of the human-induced causes of climate change, it is said that the energy sector is responsible for about three-quarters of the carbon dioxide (CO_2) emissions in the world (IPCC, 2001). The construction industry is said to be not too far behind, as it is one of the most energy intensive industries. According to the American Institute of Architects (AIA, 2000), the biggest source of emissions and energy consumption both in the U.S. and around the globe is said to be the construction industry and the energy it consumes each year. According to a briefing note prepared for the International Investors Group on Climate Change (Kruse, 2004), the cement sector alone is said to account for 5 percent of global man-made CO_2 emissions. Further, mining and manufacturing of raw materials used in construction and the transportation of heavy building materials are said to be contributing significantly to climate change.

In heavily urbanized Singapore, where half the total land area is built up, the construction sector is said to account for approximately 16 percent of national emissions of CO_2 (MEWR, 2008). This figure, however, includes emissions resulting from primary and secondary energy consumption by buildings and excludes the emissions from the industrial process involved in the construction industry. Hence, the total of direct and indirect emissions resulting from the construction industry could be much higher.

In the circumstances, it is not difficult to argue that the construction industry is one of the major industries responsible for high levels of GHG emissions, which cause climate change. Thus, in many jurisdictions, the construction sector has been identified as one of the key industries that should be pro-active in engaging in sustainable development by curtailing the GHG emissions and adopting more sustainable energy consumption patterns.

It should also be pointed out that, in addition to being energy efficient and helping mitigate climate change, the construction industry has another role to play in sustainable development. That is, improving the capacity to adapt to changing climate conditions and the related risks. For example, weather-related impacts such as hurricanes, flooding, and coastal erosion for which climate change is at least partially responsible, would require the use of new building techniques and materials to withstand adverse weather conditions. Such events would also influence the choice of site for construction projects.

Some authors (Mimura *et al.*, 1998) note that cost increases for disaster rehabilitation and countermeasures against natural calamities caused by climate change could expand the market for the construction industry. However, the flip side is that flooding and other extreme weather-related events mentioned above could damage buildings and infrastructure and cost countries billions of dollars in repair and reconstruction work. This would require diversion of precious funds earmarked for other development activities. Thus, whilst the GHG emissions by the construction industry are one of the major causes for climate change, the construction industry could also be one of the worst affected industries due to impacts of climate change.

Due to its geographic location, South East Asia has been identified as one of the most vulnerable regions to environmental risks associated with climate change. Further, the rising sea levels and the consequent flooding from the sea may also adversely affect the coastal areas in this region (IPCC, 2007). In the circumstances, it is important that the countries in the South East Asian region identify the climate change-related environmental risks to the construction industry and take initiatives to mitigate such risks.

Being a small island city state off the southern tip of the Malay Peninsula, with only 710.2 km² of land, Singapore is the smallest nation in South East Asia. However, with approximately 5 million people living in the small island, including approximately 2 million foreign workers, Singapore can be considered a densely populated, compact, cosmopolitan city. With its population expected to grow to 6 million by 2020, the construction sector in Singapore is expected to boom. Thus, the introduction of energy-efficient technologies and environmentally friendly designs and construction methods as measures to mitigate climate change whilst trying to keep the construction cost affordable to the end-users, will be one of the biggest challenges Singapore will face in the coming years. Singapore will be one of the countries in South East Asia constantly under threat due to sea erosion and rising sea levels. Thus, in addition to taking measures to reduce the GHG emissions from the construction sector, another challenge for Singapore will be to improve the adaptability of its construction industry to climate change.

12.2 Dealing with climate change and the impact on the construction industry

12.2.1 Kyoto Protocol

According to the United Nations Environmental Programme (UNEP), currently over 200 international environmental agreements and a large number of bilateral agreements on the subject of environment exist. However, the Kyoto Protocol, a protocol to the United Nations Framework Convention on Climate Change (UNFCCC)[2] which was agreed on 11 December 1997, is the only multilateral framework we have to address climate change.

The Kyoto Protocol establishes emission reduction targets for the period 2008–2012 for the industrialized countries (Annex 1 countries).[3] As far as developing countries are concerned, there are no reduction targets. However, developing countries are encouraged under the clean development mechanism (CDM),[4] one of the three mechanisms introduced to deal with global climate change,[5] to benefit from transfer of technology, and foreign investments flows into sectors such as renewable energy and afforestation projects which would contribute to mitigate GHG emissions. However, this mechanism appears ineffective as it does not monitor fair distribution of CDM projects among the developing countries (Gunawansa, 2009).

The Kyoto Protocol could be criticized for having only a limited life span and also for having binding obligations only upon the Annex 1 countries. Further, as far as adaptation is concerned, the Kyoto Protocol has not established any effective mechanism, although it requires countries to formulate, implement, publish and regularly update, national and, where appropriate, regional programmes containing measures to mitigate climate change and measures to facilitate adequate adaptation to climate change.[6] Further, although, the Kyoto Protocol has established an adaptation fund to finance concrete adaptation projects and programmes in developing countries,[7] it is ineffective, as it does not have specifically quantitative funding commitments for the countries.

The Kyoto Protocol does not contain any specific provisions directly impacting on the construction industry. However, it is likely that various legislative and policy instruments the Annex I countries introduce to meet their mandatory GHG reduction targets during the period 2008–2012, and the self-induced initiatives that non-Annex 1 countries might take to reduce the emission of GHGs, might have a direct or indirect impact on the construction industry in their individual jurisdictions, as it is one of the most energy-intensive industries.

12.2.2 National initiatives

Despite the lack of effective multilateral consensus on the responsibility of each state to deal with climate change, since the Earth Summit in 1992,[8] several nations of the world have been working towards reducing the emission of GHGs at the national level. The examples given below from the U.S., Australia and Singapore prove this point. Some of these initiatives take the form of specific legislation aimed at imposing penalties and taxation to force people and industries to adapt climate-friendly behavioural patterns and to promote sustainable development. There are other policy initiatives without specific legislation to back them. There are also voluntary industrial standards that have been introduced.

It is important to note that although the U.S. has so far not agreed to meet any mandatory GHG reduction targets under the Kyoto Protocol, 28 U.S. states have programmes to curb CO_2 emissions and at least 166 U.S. cities have agreed to apply the Kyoto emission reduction standards to their communities (Prato and Fagre, 2006). For example, in August 2006, the California State Legislature passed the Global Warming Solutions Act (AB32) which is designed to reduce the State's impact on global warming. AB32 requires a reduction of statewide GHG emissions to 1990 levels, roughly a 25 percent reduction under business-as-usual estimates, by the year 2020, using a mandatory statewide cap on emissions beginning in 2012.

The state of Wisconsin has mandated a 30 percent improvement in the energy efficiency of state buildings.[9] Some other initiatives are under consideration. One such initiative concerns the proposal for the adoption of the International Energy Conservation Code (IECC) for commercial buildings, which is projected to increase the energy efficiency of new buildings by 30 percent. It is said that the overall objective of these measures targeting buildings is to achieve zero energy usage for new residential and commercial buildings by 2030.[10]

Other states too have since followed suit, passing similar legislation. For example, the states of Oregon, New Jersey and Hawaii passed legislation in 2007 that imposed mandatory caps on GHG emissions. In Florida, the Governor has signed three executive orders in July of 2007 that set GHG emissions reduction targets for the State (PEW Centre, 2007). These instruments will have a definite impact on the construction industry due to the high energy-intensive nature of the sector.

Australia ratified the Kyoto Protocol only in December 2007. Thus, any efforts taken by Australia to formally meet its obligations under the Kyoto Protocol are still at an infant stage. According to Australia's Department of Climate Change, the country has set a long-term target for national emissions reductions of 60 per cent on 2000 levels by 2050. The building and construction sector is a key focus area for Australia's climate change response. This is because the energy use in the commercial building sector is said to generate almost 10 percent of Australia's GHG emissions according to a baseline study carried out by the Australian Greenhouse Office (Australian Greenhouse Office, 2005). The same source suggests that emissions could double between 1990 and 2010.

There are numerous environment rating schemes for buildings across Australia, that have been developed by local, state and federal governments, even though specific legislation is yet to be introduced. National Australian Built Environment Rating System (NABERS) is one example of a voluntary environmental rating system for office premises that has been developed by the Australian government. It is aimed at existing buildings and covers a range of environmental issues such as GHGs, water, stormwater, transport, landscape diversity, waste and toxic materials. It measures actual performance against set benchmarks.[11] It is tailored for use by building owners, managers and building occupants. Building owners and managers will be able to report on those aspects of the environmental performance of the building that are in their control, for example landlord energy use (lifts, air conditioning etc.) and water consumption. Building occupants will report on the environmental performance of the aspects of the building that they control (light and power in their tenancy, transport to and from the building etc.).

The Australian Building Greenhouse Rating Scheme (ABGRS), which rates the GHG performance of new and existing office buildings against benchmarks is another example. Separate ratings can apply to tenancies, core buildings or whole buildings. Actual performance must be demonstrated once the building is operating. The scheme requires buildings to report their actual annual energy use and GHG emissions, and buildings (and/or their tenancies) can only keep their ABGR star ratings if they continue to meet their target consumption levels. The rating system allows the developers, owners and occupants of office buildings to rate their greenhouse performance on a simple scale of one to five stars. Buildings that perform better receive more stars. A good rating can be a significant advantage in the commercial property market. ABGRS is now a part of NABERS and is known as NABERS Energy.[12]

Another example from Australia is the 'Green Star' rating system developed by the Green Building Council of Australia (GBCA). It is a national environmental rating tool for buildings which rates a building in relation to its management, the health and wellbeing of its occupants, accessibility to public transport, water use, energy consumption, the embodied energy of its materials, land use and pollution. Green Star aims to assist the building industry in its transition to sustainable development.[13]

In Singapore, the Green Mark Scheme introduced by the government, which will be discussed in more detail later, has introduced a mandatory certification system for buildings to establish national sustainability requirements in the construction sector.

12.3 The case of Singapore

12.3.1 Key sustainable development challenges

As noted in the introduction, Singapore is a small island city-state with limited land space. However, although it is the smallest nation in South East

Asia, in terms of gross domestic product (GDP) purchasing power parity (PPP) per capita,[14] it is currently considered as the fifth wealthiest country in the world according to the International Monetary Fund (IMF)[15] and the eighth wealthiest according to the U.S. Central Intelligence Agency's *World Fact Book* for 2009.[16] According to the government statistics, as of January 2010, Singapore's official reserves stood at US$189.6 billion.[17] Thus from an economic point of view, Singapore may not face sustainable development challenges faced by developing countries in the region.

According to Singapore's Minister for National Development Mah Bow Tan, 'We want to develop without squandering resources and causing unnecessary waste. We want to develop without polluting our environment. We want to develop, while preserving greenery, waterways, and our natural heritage' (Gov Monitor, 2010b). Thus Singapore's overall goal seems to be to grow in an efficient, clean, and green way. One of the biggest development challenges Singapore will face in achieving this is its lack of natural resources, having to rely on other countries and foreign trade for resources such as water,[18] fuel and food. Malaysia is presently supplying around 40 percent of Singapore's water needs. With approximately 2,440,000 barrels of oil imported per day, Singapore is currently ranked the seventh largest importer of oil.[19]

The country's average wind speeds are considered to be too low for the economical use of large wind turbines. The use of wave, tidal and ocean thermal resources have been ruled out due to their limited application and as the available sea space is used for ports, anchorage and shipping lanes. Singapore's geography also does not present opportunities to harness renewable energy from hydro or geothermal technologies (MEWR, 2008). Thus, the island will have to focus on raising efficiency of the current energy usage. At the same time, research and development of alternative energy sources will have to be financed.

Likewise, given Singapore's lack of water resources and the current reliance on imported water for consumption, the country will have to take initiatives to use water more conservatively by reducing the current level of consumption. Such initiatives will have to be supported with investment in infrastructure for waste water recycling and desalination of sea water.

Another key challenge is that, with no natural hinterland, Singapore has to accommodate homes, offices, industrial premises, infrastructure for utilities, parks, roads, reservoirs, airport, military facilities and many others within its small land area, whilst ensuring that a clean, green and comfortable environment to live in is provided for the people. The answer, as far as the building and construction sector is concerned, is the development of high-rise public housing, office complexes and administrative buildings in small spaces. This would, however, lead to congested and environmentally unsustainable conditions if adequate measures are not taken to develop sustainable living and working environments.

12.3.2 Construction sector in Singapore

Except for the ripple effect of the current global economic recession, it could be said that Singapore has experienced a construction boom during the last two decades, having gone through a lean period during the 1997 Asian financial crisis, followed by the terrorism threat in the region and the SARS outbreak (Chan and Gunawansa, 2008). As a result, construction is taking place in all sectors of the economy. Projects range from domestic housing and office towers through to sophisticated infrastructure developments and construction of massive resorts, hotels and other tourist attractions.

According to the Singapore government, the construction experts of BIS Shrapnel have claimed that, of the countries in Southeast Asia, the boom market will mostly benefit countries such as Singapore and Vietnam due to the rising activity in each of their domestic residential building sectors (EBIS, 2007). The rising property prices in Singapore serve as a proxy of investors' confidence in the business conditions in Singapore.

According to the Government of Singapore, in 2009, the construction sector enjoyed strong double-digit growth for the third consecutive year, driven by exceptionally strong construction demand in the preceding two years (Gov Monitor, 2010a). As a result, the industry achieved a record level of on-site construction activity or output of about $30 billion. However, the global financial crisis had a considerable impact on the construction industry in 2009, especially on private sector demand. This is evidenced by the drop in the value of contracts awarded in 2009 to $21 billion from a record high of $35.7 billion in 2008. Out of this 64 percent were public sector contracts which establish the need for public sector housing and other infrastructure facilities to cater for the growing population.

In the circumstances, a major challenge in connection with the construction sector would be the need to balance the growth of the industry with the need to mitigate the emission of GHGs and improve the adaptive capacity of buildings and infrastructure facilities to adverse impacts of climate change.

12.3.3 Commitment to reduce GHG emissions

Having ratified the Kyoto Protocol in April 2006, Singapore has made a voluntary commitment to reduce its carbon intensity by 25 percent from 1990 levels by the year 2012, although it does not have to meet any mandatory GHG reduction targets during the first commitment period. According to the Singapore Ministry of Environment and Water Resources (MEWR), in fact, the country had achieved a 22 percent reduction in 2004 (MEWR, 2006). Further, according to figures from the National Environment Agency (NEA), Singapore's carbon intensity was at 0.28 kilotonnes per SGD million in 1990 (NEA, 2002). That figure has declined to 0.21 in 2005, representing a 25 percent reduction. Thus, it could be said that Singapore is well on track to meet its reduction goals by 2012.

According to MEWR, the main contribution to Singapore's GHG emissions is CO_2. Table 12.1 shows CO_2 emissions in Singapore from the five key sectors, namely, power generation, manufacturing, transport, buildings and households, in terms of both primary and secondary consumption in 2004 (primary users are those that combust fuel directly while secondary users are those that use the electricity generated from fuel).

A clear indication from the Table 12.1 is that when the primary and secondary usage is put together, the building sector is one of the main contributors to GHG emissions in Singapore.

12.4 Singapore's framework for environmental management

Until the late 1990s, the laws on environment that existed in Singapore were diverse and scattered throughout many Acts. Further, state agencies in charge of environment-related subjects were many. However, this problem has been resolved to a considerable extent with the enactment of two key statutes, namely, the Environmental Public Health Act (EPHA),[20] which controls waste including toxic, domestic and industrial waste, and the Environmental Protection and Management Act (EPMA)[21] which consolidated various laws including the Clean Air Act and the Water Pollution Control and Drainage Act. It applies to air, water, noise, land contamination and hazardous substances.

In 2002, Singapore established the National Environmental Agency (NEA).[22] NEA's powers include inter alia:

> prescribe and implement regulatory policies, strategies, measures, standards or any other requirements on any matter related to or connected with environmental health, environmental protection, radiation control, resource conservation, waste minimization, waste recycling, waste collection and disposal and such other subject matter as may be necessary for the performance of the functions of the Agency.[23]

Table 12.1 Key CO_2 contributors in 2005

Sector	Primary consumption	Secondary consumption	Overall consumption
Electricity generation	19,058 (48%)	–	
Manufacturing	13,179 (33%)	8,311 (44%)	21,490
Transport	6,758 (17%)	921 (5%)	7,679
Buildings	391 (1%)	5,777 (30%)	6,168
Consumers/households	233 (–1%)	3,440 (18%)	3,673
Others	–	610 (3%)	610

Source: Adapted from MEWR, 2006.

Today, Singapore's framework for environmental management is based on five fundamental principles, namely, control of pollution at source; pre-emption and taking of early action; the "polluter pays" principle; innovation and new technology; and environmental ownership. According to Singapore's Ministry of Environment and Water Resources (MEWR) its mission is to 'to deliver and sustain a clean and healthy environment and water resources for all in Singapore.'[24]

The legislation mentioned above, although dealing with the subject of environment, was not enacted with the aim of dealing with climate change impact. Singapore has not passed any specific laws designed to reduce GHG emissions since ratifying the Kyoto Protocol. However, a strong national strategy has been put in place and promising industrial standards have been adopted under existing regulations to help the country achieve its emission reduction targets. In addition, several schemes have been introduced to recognize green buildings and energy efficiency, thereby, promoting cleaner construction.

12.5 National climate change strategy

According to the Singapore National Climate Change Strategy (NCCS), the country is committed to addressing climate change in an environmentally sustainable manner that is compatible with its economic growth (MEWR, 2008). The NCCS provides that the country should adopt the following guiding principles:

- Climate change action must be environmentally sustainable and compatible with the country's economic growth.
- Climate change action needs individual, corporate and government effort, as meeting the challenge of climate change cannot be solely a government initiative. Thus, in addition to bringing together representatives from various government agencies, industry representatives, academia and non-governmental organizations under the National Climate Change Committee (NCCC) for collective efforts on climate change, citizens should be motivated to join the national effort and take actions in their daily lives, whether at work, at play, or at home, to become more energy efficient and choose cleaner fuels.
- As climate change actions cover many sectors of the economy and society, the National Climate Change Strategy should be developed through a consultative, multi-stakeholder approach, taking into consideration the views of stakeholders and the public at large.

The NCCS aims to meet the national carbon intensity target through actions in five key sectors, namely, power generation, manufacturing, transport, buildings and households. The key actions that have been already taken and the planned actions for the future, which are relevant to the construction sector are briefly discussed below.

12.6 Minimum energy efficiency standards

In Singapore, the Building Control Act (Cap 29) is a prescriptive code. It prescribes standards of safety and good building practice. Accordingly, an application for approval of the plans of any building works shall be made to the Commissioner of Building Control (CoBC) by the developer of those building works.[25] Thus, except in the case of works to temporary buildings or the occupation of any such building, retrofitting of exterior features referred to in Part III of the Act, building works that are exempted under section 30 of the Act, or are in relation to a building that is so exempted, and building works that are prescribed in the building regulations to be insignificant building works, no construction can take effect without the approval of the CoBC.[26]

Section 49 of the Act provides *inter alia* that the Minister may make regulations for carrying out the purposes of the Act and for any matter which is required under this Act to be prescribed. One of the areas in which the Act enables the minister to make regulations is 'environmental sustainability measures that improve the total quality of life and minimise adverse effects to the environment, both now and in the future.'[27]

Given Singapore's tropical climate, the need for air-conditioning forms a large part of the island's electrical demand. It is expected that rising temperatures and the growing population will increase the demand for cooling. Thus, in order to promote buildings designed to encourage greater use of natural light and ventilation, and with proper insulation that ensures less energy is used to cool down buildings, Singapore has established minimum energy efficiency standards under the Building Control Regulations that focus on heat transfer. Accordingly, the Building Control Regulations 2003 provide inter alia:

> A building shall be designed and constructed with energy conservation measures to reduce –
> (a) solar heat gain through the roof;
> (b) solar heat gain through the building envelope;
> (c) air leakage through doors, windows and other openings on the building envelope;
> (d) energy consumption of lighting, air-conditioning and mechanical ventilation systems; and
> (e) energy wastage through adequate provisions of switching means.[28]

It also provides that commercial buildings with an aggregate floor area of more than 500 m^2 shall be installed or equipped with means to facilitate the collection of energy consumption data.[29]

The above regulations also require air-conditioning equipment and lighting to comply with minimum efficiency standards prescribed in the Singapore Standard Codes of Practice for Building Services and Equipment.[30] This standard provides minimum energy efficiency requirements for new installation and replacement of systems and equipment. It is used by the Building and Construction Authority of Singapore (BCA) as a reference standard in its Building Control Regulations for air-conditioning systems which exceed 30 kW cooling capacity and the maximum lighting power budget. The use of SS 530 can help businesses reduce energy consumption by 10–30 percent. This translates into significant savings in utilities. The other such codes in use in Singapore are:

- SSCP13: 1999 – Code of Practice for Mechanical Ventilation and Air-conditioning in Building;
- SSCP24: 1999 – Code of Practice for Energy Efficiency Standard for Building Services and Equipment; and
- SSCP38: 1999 – Code of Practice for Artificial Lightings in Building.

12.6.1 Building labelling scheme

To encourage best practices beyond those specified in the standard building codes, two building labelling schemes have been introduced by BCA. The first of these schemes is the 'Green Mark' scheme (GMS) introduced in January 2005 as a voluntary mechanism, which was later made mandatory by way of regulation.[31] It rates the environmental friendliness of a building based on a point scoring approach. Depending on the score, the rating is categorized in four levels, namely, 'Platinum', 'Gold Plus', 'Gold' and 'Certified'. It enables the benchmarking of the building's environmental performance and allows comparison between buildings. According to BCA, the GMS is a benchmarking scheme which aims to achieve a sustainable built environment by incorporating best practices in environmental design and construction, and the adoption of green building technologies (BCA, 2007).

The GMS has assessment criteria for two main categories, new buildings and existing buildings. The scheme for new buildings provide an opportunity for developers to design and construct green, sustainable buildings which can promote energy savings, water savings, healthier indoor environments and adoption of greenery for their projects. The scheme for existing buildings enables building owners and operators to meet their sustainable operations goals and to reduce adverse impacts of their buildings on the environment and occupant health over their entire life cycle. Points are awarded for incorporating environment-friendly features which are better than normal practice. Buildings are awarded 'Platinum', 'Gold PLUS', and 'Gold' or 'Certified' rating depending on the points scored. A minimum of 50 points is required to achieve the 'Certified' rating.[32]

12.6.2 Building Control (Environmental Sustainability) Regulations 2008

As noted earlier, the law relating to buildings in Singapore and matters connected therewith is contained in the Building Control Act.[33] Under the enabling provision that the Minister may make regulations,[34] a new regulation, Building Control (Environmental Sustainability) Regulations 2008, which came into operation in April 2008, has been introduced to promote environmental sustainability in the construction sector by introducing mandatory provisions to ensure minimum environmental sustainability standards of buildings.

These regulations apply to:

- All new building works with gross floor area of 2000 m² or more;
- Additions or extensions to existing buildings which involve increasing gross floor area of the existing buildings by 2000 m² or more;
- Building works which involve major retrofitting[35] to existing buildings with existing gross floor area of 2000 m² or more.

Under the new regulations, buildings are awarded the BCA Green Mark based on five key criteria:

- energy efficiency
- water efficiency
- site/project development and management (building management and operation for existing buildings)
- good indoor environmental quality and environmental protection
- innovation

Under the Green Mark assessment system, points are awarded for incorporating environmentally friendly features into buildings. The total number of points obtained indicates the environmental-friendliness of the building design. With the introduction of the Building Control (Environmental Sustainability) Regulations 2008, the voluntary GMS that previously existed in Singapore for both new and existing buildings, has now become mandatory except for alteration to existing buildings which does not involve major retrofitting works. Thus, it is mandatory that building works that do not fall under this exclusion achieve the minimum sustainability standard, i.e. the "Certified" rating under the GMS. The minimum Green Mark score for any building works to which the regulations apply is 50 points.

The above regulations refer to the Code for Environmental Sustainability of Buildings,[36] which establish environmentally friendly practices for the planning, design and construction of buildings which would help to mitigate the environmental impact of built structures. This Code sets out the minimum environmental sustainability standard for buildings and the administrative requirements concerning the compliance method in assessing the environmental performance of a building development.

The Code for Environmental Sustainability of Buildings provides *inter alia* that the developer or building owner shall engage the appropriate practitioners which include the qualified person (QP)[37] to ensure that the building works are designed with physical features or amenities, and may be carried out using methods and materials to meet the minimum environmental sustainability standard stipulated in Building Control (Environmental Sustainability) Regulations.[38] The QP who submits the building plan shall be overall responsible for ensuring that the minimum environmental sustainability standard is met. The QP, together with the other appropriate practitioners (i.e. PE (Mechanical) and PE (Electrical)), shall be responsible for assessing and scoring the building works under their charge. The areas of responsibility are as prescribed in Annex A of the Code.

The regulation provides that anyone who contravenes the regulations concerning meeting the minimum environmental sustainability standards[39] or deviates from the approved building plans[40] shall be guilty of an offence and shall be liable on conviction to a fine not exceeding $10,000.[41]

12.6.3 Energy Smart Labelling Scheme

Developed by the Energy Sustainability Unit of the National University of Singapore (NUS) and the National Environmental Agency (NEA), the Energy Smart Labelling Scheme (ESLS) was launched in December 2005. It is the first energy-efficient building label in Asia. Under this scheme 'Energy Smart' labelling is given to the top 25 percent most energy-efficient buildings in Singapore that also demonstrate good indoor environmental quality. It recognizes developers and owners who design and maintain efficient buildings and is also a benchmarking scheme, where building owners can compare the energy efficiency of their buildings against a national benchmark. It is a prerequisite for an existing commercial building to achieve the Energy Smart label in order to win the Platinum Award under the GMS.[42]

12.6.4 Other initiatives

In addition to the various initiatives discussed above, some of the other mechanisms that have been put in place in Singapore to help it achieve its GHG emission reduction targets include:

* Building Energy Efficiency Master Plan (BEEMP), formulated by the BCA. This details the various initiatives taken by the BCA to fulfil a number of recommendations made by the Inter-Agency Committee on Energy Efficiency (IACEE), which comprises senior officers from various government agencies (formed in 1998). The recommendations are on strategic directions to improve the energy efficiency of the building,

industry and transport sectors. Carrying out energy audits of selected buildings, review and update of energy standards, introduction of energy efficiency indices and performance bench marks and the introduction of performance contracting (also known as 'third party financing' or 'contract energy management' as a means of raising money for investments in energy efficiency based on future savings), are among the key recommendations made by the IACEE, which are implemented under the BEEMP. The plan contains programmes and measures that span the whole life cycle of a building. It promotes the use of a set of energy efficiency standards to ensure buildings are designed right from the start and continue with a programme of energy management to ensure their operating efficiency is maintained throughout their life span.[43]

- Energy Efficiency Improvement Assistance Scheme (EEIAS) is a co-funding scheme administered by the NEA to incentivize companies in the manufacturing and building sectors to carry out detailed studies on their energy consumption and identify potential areas for energy efficiency improvement.[44]
- Recently, the Government of Singapore has launched a Green Building Masterplan to encourage more developers and owners of property to go green. This has set aside S$20 million over the three years, 2008–2010, for a new scheme called the Green Mark Incentive Scheme. It is said that a further S$50 million will be set aside thereafter over the next five years to intensify research and development efforts in green building technologies and energy efficiency for buildings, particularly in the tropics (MND, 2007).

12.7 Climate change adaptation initiatives in Singapore

12.7.1 The need for adaptation initiatives

In its fourth assessment report, the Intergovernmental Panel on Climate Change (IPCC) defines adaptation as "adjustment in natural or human systems in response to actual or expected climatic stimuli or their effects, which moderates harm or exploits beneficial opportunities" (IPCC, 2007). Whilst mitigation of climate change by reducing the current levels of GHGs will help future generations, adaptation initiatives are necessary to prepare the current generation for changing and unavoidable climatic conditions.

As far as the construction industry is concerned, the recent initiatives to deal with climate change have focused on the environmental performance of buildings and construction activity, particularly emissions from buildings. Thus, as noted earlier, the focus of these initiatives is on mitigation. Given that there is no scientific evidence establishing that climate change could be completely reversed, there is a need to anticipate and deal with the consequences of changing climate conditions, while at the same time working to achieve long-term reductions in GHG emissions. In the circumstances, what is needed is a two-pronged system of initiatives, namely:

- Initiatives to mitigate climate change by reduction of GHG emissions; and
- Initiatives to improve the adaptation capacity to climate change impacts.

The initiatives that fall under the first category above are clearly visible in most countries. However, as far as the second is concerned, very little has been done. Lack of knowledge on the measures to be taken and lack of coordination and support among relevant stakeholders, could be identified as the key reasons. Further, not every country in the world has the technical and financial capacity to take adequate adaptation initiatives to deal with climate change.

It is important for Singapore to understand that mitigation and adaptation are not alternatives and that both need to be pursued actively and in parallel. This is because whilst mitigation of climate change is essential, adaptation to changing climate conditions is inevitable (UNFCC, 2006). Mitigation is essential because, without firm action now, future generations could be confronted with climate change on a scale so overwhelming that adaptation might no longer be feasible. Mitigation will not be enough on its own. This is because even if current efforts to reduce GHG emissions are successful, some adaptation will be inevitable since climate change occurs only after a long time-lag. Just as the current global warming is the consequence of emissions decades ago, the process will continue. Thus, even the most rigorous mitigation efforts now might not be able to prevent climate changes in the near future.

As noted earlier, Singapore seems to have taken several effective measures towards mitigation of climate change. The fact that Singapore is well on target in achieving its self-imposed commitment to reduce GHG emissions indicate that the efforts taken so far by the country towards climate change mitigation have been successful. Being a tiny island state, Singapore could be vulnerable to the adverse effects of climate change. Thus, it is important that the government of Singapore considers the implementation of effective measures aimed at climate change adaptation. Having a sound economy with an easily manageable landmass and population, Singapore is in an ideal position to take proactive measures, not only towards mitigation of climate change, but, also towards adaptation to climate change.

12.7.2 Current adaptation initiatives

In the NCSS the NEA has identified the following seven potential impacts of climate change (MEWR, 2008): increased flooding, coastal land loss, water resource scarcity, public health impact from resurgence of diseases, heat stress, increased energy demand and impacts on biodiversity. Of the adaptation policies to address these impacts, those which are relevant to the construction industry are briefly discussed below.

In the early 1970s, many areas in Singapore were prone to flooding during the monsoon months. During the last 30 years, the Public Utilities Board (PUB) has implemented infrastructure policies to reduce flood-prone areas. The result is the reduction of the flood prone area from 3200 ha in the 1970s to about 124 ha in 2006 (MEWR, 2008). According to NCCS, the aim is to further reduce the area to less than 66 ha by 2011. This is expected to help ameliorate inland flooding incurred by either sea-level rises (that may make it more difficult for rainwater to drain back into the sea) or storm surges at the shoreline caused by strong winds blowing inward from the sea.

Singapore has a relatively flat coastline and being land-strapped means that the coasts of Singapore are well utilized for a range of purposes, including recreational facilities and residential buildings. According to NCCS, currently, about 70–80 percent of Singapore's coastal areas are protected by hard wall or stone embankments; the remaining 20–30 percent are either natural areas such as beaches and mangroves (MEWR, 2008). It is believed that a sea level rise of up to 59 cm can result in some coastal erosion and land loss in Singapore. Thus, the NEA and MEWR are looking into ways to strengthen and protect vulnerable regions.

Singapore has been successful in facing up to its limited water resources with a series of integrated water policies. Rising global temperatures may result in changes in rainfall patterns and thus reduce the amount of water captured and stored in reservoirs. According to NCCS, diversification of its water sources, a key component of Singapore's current water policy, will be helpful in tackling the likely rainfall reduction due to climate change. Singapore has been successful, and is improving, in the recycling of used water to produce water that is safe enough to consume (also known as NEWater); it is also developing desalination. However, if Singapore is to reduce its current reliance on imported water, more infrastructure facilities in the water sector will have to be developed.

NCCS notes that increase in global temperature will have profound effect on Singapore, which is a tropical island. For example, extreme heat might result in the increase of air-conditioning usage in the island (MEWR, 2008). Thus, it notes that innovative measures are needed to reduce or change the use of air-conditioners in buildings and facilities, as they are one of the most energy-intensive mechanical systems utilized in cities. One way is to ensure that renewable energy resources are used to power the utilities; another approach is to increase the energy efficiency of these mechanical systems, either through a change in the design of the technology or improvement in the operation and maintenance of the systems. In recent years, Singapore has begun to explore more extensive use of passive methods to reduce the heat island effect and the need to use mechanical ventilation systems. An example is the deployment of urban greenery in the city and modification of building layouts and designs to reduce the cooling load of buildings (for example, through the use of building materials with better thermal properties and lighter-coloured building surfaces). The Urban Redevelopment Authority

(URA), Housing and Development Board (HDB) and National Park Board (NPB) have come together to help promote the use of city parks and rooftop and vertical greenery to reduce local heat stress (MEWR, 2008). Together with the GMS, these strategies are expected to serve to reduce the overall cooling load (hence, energy consumption) of buildings.

Singapore's public health and medical policies have always addressed the control of tropical vector-borne diseases, particularly dengue fever. Dengue patterns are affected by many factors, including climate. According to NCCS, NEA is currently studying the link between climatic factors such as temperature, humidity and rainfall with dengue cases. Further, the government has put in place a comprehensive mosquito surveillance, control and enforcement system, which includes a review of building designs to reduce potential breeding habitats, including forbidding the use of roof gutters in new buildings except in special circumstances.

The construction industry has been identified by the Singapore government as a key sector in which potential climate change impacts such as increased flooding, coastal land loss, heat stress and increased energy consumption can be effectively addressed (MEWR, 2008). By improving the drainage infrastructure, the Public Utilities Board (PUB) aims to reduce the local flood-prone area to less than 66 ha by 2011. Further, since 1991, PUB requires new reclamation projects to be built to a level 125 cm above the highest recorded tide level, higher than the average expected sea level rise of 59 cm due to climate change. Nonetheless, as this 59 cm rise will still affect the coastal areas, according to MEWR (2008), the government is considering strengthening and increasing the height of the present stone embankment to control the degree of coastal erosion.

12.7.3 Future adaptation initiatives

In February 2008 the Singapore government set up an Inter-Ministerial Committee on Sustainable Development (IMCSD) to formulate a clear national framework and strategy for Singapore's sustainable development in the context of the emerging domestic and global challenges. The committee was co-chaired by the Minister for National Development and the Minister for the Environment and Water Resources and consisted of three other ministers in charge of the subjects of Finance, Transport and Trade and Industry respectively (MEWR, 2009a). In April 2009 the IMCSD released the Sustainable Singapore blueprint report, 'A Lively and Liveable Singapore: Strategies for Sustainable Growth' (Singapore Blueprint). The strategies in the report were based on the rationale of 'The Singapore Way', which is described as follows (MEWR, 2009b):

> For Singapore, sustainable development means achieving both a more dynamic economy and a better quality living environment, for Singaporeans now and in the future.

We need the economy to grow. This creates jobs, raises our standard of living, and yields the resources that we need to safeguard our environment. But we must grow in a sustainable way, or else a high GDP per capita will be achieved at the expense of our overall quality of life, and cannot be maintained over the longer term. Protecting our environment safeguards a high standard of public health for our people, and makes our city attractive to Singaporeans and foreigners alike.

The IMCSD details several key goals and initiatives for the period from now until 2030 to improve resource efficiency and enhance Singapore's urban environment. Some of the key initiatives that are of particular relevance to the construction industry and the built environment are listed below:

- Reduction of the energy intensity (per dollar GDP) by 35 percent from 2005 levels by 2030;
- Targeting 80 percent of the existing building stock (by GFA) to achieve at least Green Mark Certified rating (minimum level of energy efficiency) by 2030;
- Introduction of solar technology at 30 public housing precincts nation-wide as a pilot project;
- The Housing Development Board (HDB) to reduce energy use of common areas of public housing by 20–30 percent and build more eco-friendly HDB housing;
- Providing 0.8 ha of park land per 1,000 persons by 2030;
- Implementation of a S$100 million Green Mark Incentive Scheme for existing buildings to undergo energy efficiency retrofitting;
- Introduction of a Green Mark GFA incentive scheme for new buildings that can attain Green Mark 'Gold Plus' and 'Platinum' rating;
- Incorporation of Green Mark 'Gold Plus' and 'Platinum' requirements will be as part of land sales requirements; and
- Improve air quality by reducing ambient PM 2.5 (fine particles) levels to an annual mean of 12 $\mu g/m^3$ and capping ambient SO_2 (sulphur dioxide) levels at an annual mean of 15 $\mu g/m^3$.

All of the above initiatives and goals set by the Singapore Blueprint are commendable as effective and efficient means by which a country with hardly any natural resources is trying to incorporate concepts of sustainability into its development and nation building policy architecture. However, being a small tropical island with a very limited land mass, Singapore should take more measures to build up resilience to adverse impacts of climate change.

12.7.4 Recommendations

Mitigation and adaptation are not alternatives and both need to be pursued actively and in parallel. An important factor to note is that adaptation cannot entirely be left to social or market forces. Essential forms of adaptation will demand that institutions, both public and private, plan their strategies and take action in advance. For example, coastal authorities will have to address the sea level rise by building dykes. Vulnerable areas such as coastal settlements will have to be identified and mandatory construction standards will have to be introduced in relation to construction works in such areas. At the same time, the private sector should also be encouraged to develop new and innovative designs that would change the established styles of architecture of buildings and contribute to adaptation to climate change. This may require the introduction of new laws and building codes that go beyond the promotion of voluntary construction standards and current GMS, which is essentially aimed at mitigation of climate change by reducing GHG emissions from buildings.

In addition, the government needs to provide the housing and commercial building sectors with appropriate information and education on adaptation through industry training schemes. In a highly competitive construction sector, those engaged in construction may be profit driven, with research and development of sustainable construction methods and materials taking low priority. Thus, there is also a need to support the construction industry through programs of research, development and demonstration. In this respect, it should be noted that the effective role played by the government institutions such as the NEA and BCA in participating in and promoting research and development of new technologies and industry practices have contributed heavily to Singapore's success in being well on track for achieving its Kyoto GHG reduction targets by the deadline of 2012. Similar leadership for adaptation is recommended.

A serious problem that should be taken into consideration is the lack of professional and contractor expertise regarding energy efficient design and construction and those that are aimed at enhancing adaptability to changing climate conditions. If there is no accountability for designers to ensure that buildings actually perform as predicted, the measures taken towards achieving the targeted mitigation levels and adaptation standards may be futile. In this connection, Singapore could learn from Australia and consider the implementation of a program like the Climate Change Adaptation Skills for Professionals Program (DEWR, 2007). This provides financial grants to tertiary education and training institutions and professional associations to develop professional development and accreditation programs geared towards architects, engineers, natural resource managers and planners so that a pool of experts with skills to deal with adaptation to climate change could be built up.

12.8 Conclusions

Climate change is an issue that can be no longer ignored. Thus, countries will have no choice but to reform and restructure their development policies to complement any initiatives taken to address climate change. With growing global concerns on the sustainability of the built environment, there is increasing pressure for the construction industry to consider the environmental impact of construction projects and the need to take into account the importance of climate change and its impact on the built environment.

It should be understood that achieving the expected level of GHG emission reduction and promoting energy efficiency and green buildings is not going to be an easy task. For example, as noted above, green buildings would require the restructuring of current forms of energy usage in the construction sector to ensure that GHG emission is reduced to the expected levels. However, as any such restructuring would require the use of more energy efficient materials and less energy consuming processes in the construction process, at least in the short term, cost increases in construction is a likely scenario.

The possibility that there will be cost increases in construction due to new initiatives to mitigate climate change may not be a long-term issue. Once new and energy efficient technologies are developed, due to their low energy consumption, the cost of construction is likely to come down. It is said that, although environmental friendly buildings will cost about 5–10 percent more up front, they will bring future saving of about 10–15 percent on energy expenses (BCA, 2007). Further, it is said that green buildings could yield up to 30 percent savings in energy consumption through green features such as building envelope designs which reduce heat absorption, provide more daylight, maximize natural ventilation, and use more energy efficient air-conditioning systems and light fittings (BCA, 2007).

In the circumstances, the policy makers and regulators in Singapore will have to take into consideration issues such as the demand for new and cheaper buildings; the cost of finding new and energy-efficient technologies; the need to meet national and international emission reduction standards; the need to restructure existing practices and energy usages in the industry to be more environmental friendly; and the need to be alert to adapt to changing climate conditions which are likely to be inevitable, when planning and taking decisions relating to building and construction industry. In other words, the entire life cycle of buildings could be changed in order to establish a coherent succession of sustainable construction activities that would be in harmony with national policies aimed at reducing GHGs and contribute to the mitigation of, and adaptation to, climate change.

The initiatives that have been taken so far in Singapore indicate that the country has achieved a lot through its strategic planning and by introducing good practices to the construction industry. Further, the effective role played by the government institutions such as the NEA and BCA in participating in and promoting research and development of new technologies and industry practices have contributed heavily to Singapore's success in being well on track for achieving its GHG reduction targets by the deadline of 2012. The initiatives such as the GMS are initiatives most countries could easily take to encourage the construction industry to consider developing green buildings. Such schemes give guidance and incentives to developers and building owners to pursue energy efficiency in buildings, in order to achieve both environmental performance and significant cost savings.

Finally, it should be said that Singapore has the prerequisites for the development of new and clean energy and construction industries. Its strong industrial base, reliable and efficient intellectual property protection regimes, excellent infrastructure, skilled labour, and support from the service sectors such as banks and insurance agencies, would provide the necessary foundation for such development. However, in order to ensure that such development would be sustainable, the policy initiatives taken will have to be backed with effective legislation designed to implement the relevant policies, regulate the relevant industry practices and punish the wrongdoers.

Notes

1 GHG refer to gases that absorb the infrared radiation from the sun and warm the lower atmosphere, such as carbon dioxide, nitrous oxide, methane, sulphur hexafluoride and hydrofluorocarbons.
2 In its Article 2, setting out its objective, the UNFCC provides:

> The ultimate objective of this Convention and any related legal instruments that the Conference of the Parties may adopt is to achieve, in accordance with the relevant provisions of the Convention, stabilization of greenhouse gas concentrations in the atmosphere at a level that would prevent dangerous anthropogenic interference with the climate system. Such a level should be achieved within a time frame sufficient to allow ecosystems to adapt naturally to climate change, to ensure that food production is not threatened and to enable economic development to proceed in a sustainable manner.

3 Article 3.
4 Article 12.
5 The other two mechanisms are: (1) Joint Implementation (Article 6) and (2) Emission Trading (Article 17).
6 Article 10(b).
7 Article 12 (8).
8 United Nations Conference on Environment and Development (UNCED), Rio de Janeiro, 3–14 June 1992.
9 Executive Order 145, 11 April 2006.
10 Ibid.
11 For more information on NABRES see: www.nabers.com.au
12 For more information on ABGRS see: www.abgr.com.au
13 For more information on Green Star see: www.gbcaus.org
14 Comparisons of national wealth are frequently made on the basis of PPP to adjust for differences in the cost of living in different countries.
15 IMF, World Economic Outlook Database, October 2008.
16 CIA, *World Fact Book* 2009, https://www.cia.gov/library/publications/the-world-factbook/rankorder/2004rank.html
17 Singapore Department of Statistics (2010), IMF Special Data Dissemination Standard – Economic and Financial Data for Singapore, http://www.singstat. gov.sg/sdds/data.html

18 Malaysia is presently supplying around 40 percent of Singapore's water needs.
19 With approximately 2,440,000 barrels of oil imported per day, Singapore is currently ranked the seventh largest importer of oil. http://www.indexmundi. com/singapore/oil_imports.html
20 Cap 95.
21 Cap 94A.
22 National Environmental Agency Act (Cap 195).
23 Section 11 of the National Environmental Agency Act.
24 See the vision and mission statements of MEWR: http://app.mewr.gov.sg
25 Section 5.
26 Section 4.
27 Section 49(2)(e)(viii).
28 Section 30.
29 Section 31.
30 SS 530:2006.
31 See Section 4.2.3 below.
32 For more information, see: www.bca.gov.sg
33 Cap. 29.
34 Sec. 49 (2) (e) (viii).
35 Major retrofitting works would include works involving major overhauling of air-conditioning system and significant modification to building façade along with other building services.
36 The Code for Environmental Sustainability of Buildings, Version 1, BCA, 2008.
37 A QP is defined in Section 2 the Building Control Act (Cap 29) as a person who is registered as an architect under the Architects Act (Cap. 12) and has valid a practicing certificate issued under that Act or a professional engineer under the Professional Engineers Act (Cap. 253) and has valid practicing certificate issued under that Act.
38 Section 3.3.
39 Section 6(1).
40 Section 8(a).
41 Section 10(1).
42 For more information on the ESLS see: http://www.esu.com.sg/research2.html
43 For more information on BEEMP, see: http://www.bdg.nus.edu.sg/Building Energy/energy_masterplan/index.html
44 For more information on EEIAS see: http://app.nea.gov.sg/cms/htdocs/article. asp?pid=2536

References

American Institute of Architects (AIA) (2000) 'Architects and Climate Change', Background Paper, American Institute of Architects, New York.
Australian Greenhouse Office (2005) 'National Climate Change Adaptation Programme', Department of the Environment and Heritage, Commonwealth of Australia.
Building and Construction Authority (BCA) (2007) 'Green Buildings', an advertorial, BCA, Singapore, The Straits Times, 21 March 2007.
Chan, P., and Gunawansa, A. (2008) 'Resolving Construction Disputes the ASEAN Way', *Building and Construction Law*, 24, p 313.
Department of Environment and Water Resources (DEWR) (2007) *Climate Change Adaptation Actions for Local Government, Department of Environment and Water Resources*, Australian Greenhouse Office, Government of Australia.

EnterpriseOne Business Information Services (EBIS) (2007) EnterpriseOne Focus, October, http://www.ebis.sg/Portals/0/pdfs/EnterpriseOneFocus/EnterpriseOne Focus – Issue1007.pdf [Accessed on 14 January 2010].

Gov Monitor (2010b), "Sustainable Housing: Charting new frontiers – the Singapore Perspective", *Keynote speech by Minister for National Development Mah Bow Tan at the opening of the International Housing Conference 2010*, http://the govmonitor.com/world_news/asia/singapore-sees-sustainable-development-and-public-housing-as-priority-22469.html [Accessed on 14 February 2010].

Gov Monitor (2010a), Speech by Ms. Fu Hai Yien, Senior Minister of State for National Development, BCA-Redas Construction & Property Prospects 2010 Seminar, 13 January 2010, http://thegovmonitor.com/world_news/asia/singapore-highlights-positive-outlook-for-construction-industry-21194.html [Accessed on 14 February 2010].

Gunawansa, A., (2009) 'The Kyoto Protocol and Beyond: A South Asian Perspective', in Koh, K., *et al.*, *Crucial Issues in Climate Change and the Kyoto Protocol: Asia and the World*, World Scientific, Singapore.

Intergovernmental Panel on Climate Change (IPCC) (2001) *Working Group I: Third Assessment Report*, Cambridge University Press, Cambridge.

IPCC (2007) *Working Group I: Fourth Assessment Report*, Cambridge University Press, Cambridge.

Kruse C. (2004) 'Climate Change and the Construction Sector', Briefing Note, Institutional Investors Group on Climate Change, London.

Ministry of Environment and Water Resources (MEWR) (2006) 'Speech by Assoc Prof Koo Tsai Kee, Senior Parliamentary Secretary, at the Launch of the Solar Roof Project Singapore', News Release No. 09/2006, MEWR, Singapore.

MEWR (2006) 'National Climate Change Strategy', available at: http://app.mewr. gov.sg [Accessed on 21 January 2010].

MEWR (2008) 'Singapore's National Climate Change Strategy', available at: app. mewr.gov.sg/data/ImgUpd/NCCS_Full_Version.pdf [Accessed on 21 January 2010].

MEWR (2009a) 'The Inter-Ministerial Committee for Sustainable Development Unveils Blueprint for a Sustainable Singapore', available at: http://app.mewr.gov. sg/web/contents/ContentsSSS.aspx?ContId=1307 [Accessed on 21 January 2010].

MEWR (2009b) 'A Lively and Liveable Singapore: Strategies for Sustainable Growth', available at: http://app.mewr.gov.sg/data/ImgCont/1292/sustainbleblue print_forweb.pdf [accessed on 11 January 2010].

Mimura, N., T. Ichinose, H. Kato, J. Tsutsui, and K. Sakaki (1998) 'Impacts on Infrastructure and Socio-economic System', in Nishioka, S., and Harasawa, H., *Global Warming: The Potential Impact on Japan*, Springer-Verlag, Tokyo, Japan, pp. 165–201.

Ministry of National Development (MND) (2007) 'Greening the Built Environment', Speech by Ms Grace Fu, Singapore's Minister of State for National Development, at the BCA Green Mark Seminar, 20 March 2007, available at: http://www. mnd.gov.sg [Accessed on 11 January 2010].

National Environmental Agency (NEA) (2002) 'Energy Efficiency Improvement Assistance Scheme', available at: http://app.nea.gov.sg [Accessed on 21 January 2010].

PEW Centre on Global Climate Change (2007) 'California Global Warming Solutions Act of 2006', available at: http://www.pewclimate.org/what_s_being_ done/in_the_states/ab32/ [Accessed on 12 January 2010].

Prato, T., and Fagre, D. (2006) 'Coping with Climate Change', American Institute of Biological Sciences, Washington.

United Nations Framework Convention on Climate Change (UNFCC) (2006) *Technologies for Adaptation to Climate Change*, Bonn, Germany.

Index